Correlative Imaging

Current and future titles in the Royal Microscopical Society – John Wiley Series

Published

Principles and Practice of Variable Pressure/Environmental Scanning Electron Microscopy (VP-ESEM)
Debbie Stokes

Aberration-Corrected Analytical Electron Microscopy
Edited by Rik Brydson

Diagnostic Electron Microscopy – A Practical Guide to Interpretation and Technique
Edited by John W. Stirling, Alan Curry, and Brian Eyden

Low Voltage Electron Microscopy – Principles and Applications
Edited by David C. Bell and Natasha Erdman

Standard and Super-Resolution Bioimaging Data Analysis: A Primer
Edited by Ann Wheeler and Ricardo Henriques

Electron Beam-Specimen Interactions and Applications in Microscopy
Budhika Mendis

Biological Field Emission Scanning Electron Microscopy 2V Set
Edited by Roland Fleck and Bruno Humbel

Understanding Light Microscopy
Jeremy Sanderson

Correlative Microscopy in the Biomedical Sciences
Edited by Paul Verkade and Lucy Collinson

Forthcoming

The Preparation of Geomaterials for Microscopical Study: A Laboratory Manual
Owen Green and Jonathan Wells

Electron Energy Loss Spectroscopy
Edited by Rik Brydson and Ian MacLaren

Correlative Imaging

Focusing on the Future

Edited by

Paul Verkade
University of Bristol
Bristol
United Kingdom

Lucy Collinson
The Francis Crick Institute
London
United Kingdom

Registered Offices
John Wiley & Sons, Inc., 111 River Street, Hoboken, NJ 07030, USA
John Wiley & Sons Ltd, The Atrium, Southern Gate, Chichester, West Sussex, PO19 8SQ, UK

Editorial Office
The Atrium, Southern Gate, Chichester, West Sussex, PO19 8SQ, UK

For details of our global editorial offices, customer services, and more information about Wiley products visit us at www.wiley.com.

Wiley also publishes its books in a variety of electronic formats and by print-on-demand. Some content that appears in standard print versions of this book may not be available in other formats.

Library of Congress Cataloging-in-Publication data applied for

ISBN: 9781119086451

Cover Design: Wiley
Cover Images: Courtesy of Lucy Collinson, Chris Peddie, and Martin Jones, © Christos Georghiou/Shutterstock, © Andrii Vodolazhskyi/Shutterstock, © Choksawatdikorn/Shutterstock, © Sergey Nivens/Shutterstock

Set in 10/12pt Warnock by SPi Global, Pondicherry, India
Printed and bound in Singapore by Markono Print Media Pte Ltd

10 9 8 7 6 5 4 3 2 1

Contents

List of Contributors

Kurt Anderson
The Francis Crick Institute, London,
United Kingdom

Tanmay A.M. Bharat
Sir William Dunn School of Pathology,
University of Oxford, United Kingdom

Jose L. Carrascosa
Department of Macromolecular
Structures, Centro Nacional de
Biotecnologia (CNB-CSIC), Madrid, Spain

Francisco Javier Chichón
Department of Macromolecular
Structures, Centro Nacional de
Biotecnologia (CNB-CSIC), Madrid,
Spain

Georg Fantner
Laboratory for Bio- and Nano-
instrumentation, School of Engineering,
Interfaculty Institute of Bioengineering,
Lausanne, Switzerland

Julia Fernandez-Rodriguez
Centre for Cellular Imaging
at Sahlgrenska Academy,
University of Gothenburg, Sweden

Christopher J. Guérin
VIB Bioimaging Core, Ghent, VIB
Inflammation Research Center, Ghent
and Department of Molecular Biomedical
Research, University of Ghent, Belgium

J. P. Hoogenboom
Imaging Physics, Delft University of
Technology, The Netherlands

Eija Jokitalo
Helsinki Institute of Life Science,
Institute of Biotechnology, University of
Helsinki, Finland

Niels de Jonge
INM – Leibniz Institute for New
Materials and Department of Physics,
Saarland University, Saarbrücken,
Germany

Judith Klumperman
Section Cell Biology, Center for
Molecular Medicine, University
Medical Center Utrecht,
The Netherlands

Irina Kolotuev
University of Lausanne, EM Facility,
Switzerland

R. I. Koning
Cell and Chemical Biology, Leiden
University Medical Center, The
Netherlands

A. J. Koster
Cell and Chemical Biology, Leiden
University Medical Center, The
Netherlands

Wanda Kukulski
Cell Biology Division, MRC Laboratory of
Molecular Biology, Francis Crick Avenue,
Cambridge, United Kingdom

Frank Lafont
Cellular Microbiology and Physics of
Infection Group
Center for Infection and Immunity
of Lille,
CNRS UMR8204 – Inserm U1019 – Lille
Regional University Hospital
Center – Institut Pasteur de Lille – Univ.
Lille, France

R. I. Lane
Imaging Physics, Delft University of
Technology, The Netherlands

Saskia Lippens
BioImaging Core, VIB, Ghent,
Belgium

Nalan Liv
Section Cell Biology, Center for
Molecular Medicine, University Medical
Center Utrecht, The Netherlands

Kristina D. Micheva
Stanford University School of Medicine,
California, United States

Tommy Nilsson
The Research Institute of the McGill
University Health Centre and McGill
University, Montreal, Canada

Ardan Patwardhan
European Molecular Biology Laboratory,
European Bioinformatics Institute
(EMBL-EBI), Wellcome Genome
Campus, Hinxton, United Kingdom

Perrine Paul-Gilloteaux
Structure Fédérative de Recherche
François Bonamy, CNRS, INSERM,
Université de Nantes, France

Christopher J. Peddie
Electron Microscopy Science Technology
Platform, The Francis Crick Institute,
London, United Kingdom

Eva Pereiro
Mistral beamline, ALBA Light Source,
Cerdanyola del Vallès, Barcelona, Spain

A. Srinivasa Raja
Imaging Physics, Delft University of
Technology, The Netherlands

Nicole L. Schieber
Cell Biology and Biophysics Unit,
European Molecular Biology Laboratory,
Heidelberg, Germany

Martin Schorb
European Molecular Biology Lab (EMBL),
Heidelberg, Germany

Jason R. Swedlow
Centre for Gene Regulation and
Expression, University of Dundee, United
Kingdom

Preface

Correlative microscopy (CM), or more broadly correlative imaging (CI), aims to analyze a single sample by two or more distinct imaging modalities. By doing so, one should be able to extract more scientific insight than would have otherwise been possible using each imaging modality as a standalone technique. We have thus coined the expression 1 + 1 = 3 to explain the principle of CI. It should be noted that CI is NOT the process of imaging biological replicates with a variety of imaging techniques, which would be more properly referred to as comparative imaging.

Over the last two decades, the field of correlative imaging has seen a massive expansion in development and application, primarily driven by the need to link structure and function in a biological context. This expansion was facilitated by a number of factors, including the development of superresolution light microscopy, the resolution revolution in cryo-electron microscopy (EM), and the volume revolution in scanning electron microscopy (SEM).

The correlative revolution began with the development of correlative light electron microscopy (CLEM), with an initial swell at the end of the 1980s that then exploded in terms of developments and publications in the early 2000s (see also Chapter 2). CLEM specifically combines a light and an electron microscopy modality to image the same sample, and is the best-established CI methodology. In the early days, separate CLEM sessions in microscopy conferences would highlight technical advances in the field, and those, expanded into CI sessions, are now a mainstay at most microscopy conferences. As CI has matured, the most established workflows have shifted into the applications domain, and are often incorporated into mainstream scientific sessions at biological and, increasingly, physical sciences meetings. This important transformation shows that CI technology is now considered an established technique that can be applied to a wide variety of research questions.

Not all research questions will need a CI approach, but where the region of interest within the sample to be imaged is rare in space and/or time, CI can deliver "the needle in the haystack," alongside significant savings in both time and resources. In addition, many scientific questions will require adaptation or optimization of an existing correlative workflow, or even development of a new CI approach. To this end, we have already collected a large number of CLEM approaches and published them in dedicated volumes of the *Methods in Cell Biology* book series (Volumes 111, 124, and 140). Here, we asked the authors of the chapters to describe their technical approach, highlight tips and tricks, and, importantly, to explain why they had chosen their approach to answer their biological research question. With continuing fast-paced developments in the field, we have already received a number of queries for a fourth edition, for which we are compiling a list of chapters, with no shortage of new material available.

The feedback we have received on those books has been very positive. They are a resource that captures a snapshot of the state of the art in the field at the point of publication, and technology developers have used them as inspiration for the next iterations of new CI approaches. Having compiled the current state of the art, we were interested to look at where we are heading next. We asked leaders in different areas of CI to write down their thoughts on current limitations and how these could be solved, and what the next transformative technologies might be. Thus, this book is a snapshot of the current state of the art, but with additional musings and best-guesses of leaders in the field as to what future generations of CI technology may look like.

We asked these experts an additional question, "What CI technology would you ideally use to answer our scientific questions?" This "blue-skies daydreaming" exercise was not to be limited by the practicalities of current hardware and software solutions, and turned out to be fruitful in generating a call from the community for concerted efforts in specific areas, as well as delivering fascinating ideas that will undoubtedly drive new breakthroughs.

We fully realize that the chapters in this book are a personalized choice of topics and we may well have missed some of the next transformative CI technologies. We also recognize that some of the chapters will have a more general impact and will be valid for other research fields as well. We look forward to reading the book in 5 years' and 10 years' time, to see how the future of CI matches up to the expert predictions of 2017–19, when the book was written.

We hope you enjoy the read and that this book may be an inspiration for your own research.

February 2019

Paul Verkade
Lucy Collinson

1

It's a Small, Small World: A Brief History of Biological Correlative Microscopy

Christopher J. Guérin[1], Nalan Liv[2], and Judith Klumperman[2]

[1] VIB Bioimaging Core, Ghent, VIB Inflammation Research Center, Ghent and Department of Molecular Biomedical Research, University of Ghent, Belgium
[2] Section Cell Biology, Center for Molecular Medicine, University Medical Center Utrecht, The Netherlands

1.1 It All Began with Photons

Light microscopy (LM) is arguably the oldest technology still used in scientific research today. Until the mid-1600s, the world of structures smaller than about 400 microns was unseen and unknown. While the principles of using lenses to magnify were known as far back as Euclid (c. 300 BCE) [1], microscopy had to await technical developments in the manufacture of lenses and the casings to hold and position them, before they could be used to extend the power of human visual resolution. The earliest published description of a biological sample viewed using a simple one lens microscope was probably in 1658's Scrutinium pestis physico-medicum [2] written by a German friar Athanasius Kircher. In this manuscript he describes the presence of "little worms" in blood that he associates with disease; thus anticipating the germ theory by almost 100 years.

Around the same period, Dutchman Antonie van Leeuwenhoek used his single-lens microscope to examine samples of mold, bees, and lice, and reported these and other observations to the Royal Society in a series of letters beginning in 1673. It was when he went on to look at samples of blood, tooth plaque, and sperm that he observed that individual small structures that moved of their own volition! When he reported his observations in a letter to the Royal Society in London in 1676, they were met with great skepticism. In 1677, a delegation was sent to determine if he was brilliant or demented. Having vindicated his observations, he was elected to the Royal Society in 1680. However, while the best of van Leeuwenhoek's microscopes had an impressive maximum magnification of 260 times, their resolving power was limited to about 1.4 μm [3].

Although simple one-lens microscopes like Van Leeuwenhoek's were impressive, a Dutch inventor by the name of Cornelius Jacobszoon Drebbel brought a new device to London [4] even earlier (1619), a two-lens microscope that possessed higher magnification capacity than the Van Leeuwenhoek instrument since it was based on the principle that in a two-lens microscope the total magnification of the lenses was multiplicative [5]; although the resolution was limited by optical aberrations.

Using a microscope very much like Drebbel's, but with an improved source of illumination, the Englishman Robert Hooke was able to see details in pieces of plants, animals, and insects that had previously been unknown. For example, he observed that a piece of cork bark was composed of many small rectangular compartments. They reminded him of the small rooms that monks slept in. He called them cells, a name we still use today; had he called them chambers we might be studying chamber biology instead. He published these observations as well as the first recorded attempt to make measurements using a microscope in his 1665 book *Micrographia* [6]. These early studies of the invisible world of cells represent the birth of modern microscopy.

In the eighteenth and nineteenth centuries, microscopes became progressively more powerful, lens design was improved to remove aberrations, and innovations such as the use of polarized illumination were introduced. In the 1880s the German scientists Ernst Abbe and August Valentin Köhler working with Carl Zeiss brought together a sophisticated lens design [7] and improved illumination methods [8, 9] to create microscopes that could resolve subcellular structures. Abbe was the first to mathematically calculate the limits of microscope resolution using photons [10]. His calculations showed that the wavelength of visible light and the angle from which the diffracted light is collected defined the limits for microscopic resolution. Thus, the Abbe diffraction barrier of 188 nm was elucidated, and this would remain the limit of light microscopy until the advent of super-resolution techniques some 125 years later.

1.2   The Electron Takes Its Place

In the 1920s, while light microscopy still had to fully exploit its resolution possibilities, a young French physics student was pondering the theories of Einstein, in particular the nature of electrons, and wondering if they had a wavelength. His name was Louis de Broglie and the equation describing the wave nature of electrons was at the heart of his PhD thesis [11]. In a triumph of early career achievement his thesis secured him the 1929 Nobel prize in physics! Being a theoretician, he had no practical use for his work and went on to the next equation. Fortunately, there were more practically minded physicists who did see the use of the wave nature of electrons. Ladislaus Marton in Brussels, and Ernst Ruska, Max Knoll, and Ernst Brüche in Berlin developed simultaneous prototype transmission electron microscopes, which proved that not only did electrons have a wavelength but also that they could be focused by electromagnetic lenses and used in the same manner as light was used in optical microscopy [12]. Ruska theorized that under the right conditions these microscopes could achieve a resolution of 2Å, which was proved correct almost 40 years later [13].

Biologists rejoiced at the news that smaller subcellular structures could finally be resolved; however, it came at a price. Specimens had to be imaged in high vacuum and radiation damage from the strong electron beam was intense. Despite that, Marton published the first biological electron micrograph of a sample of Drosera intermedia, sundew, in the journal *Nature* in 1934 [14]. While this was a breakthrough, the actual resolution of electron micrographs would be insufficient to produce useful scientific data for another 20 years. So until almost the 1960s, electron microscopes were like the optical microscopes of the seventeenth century, largely curiosities.

1.3   Putting It Together, 1960s to 1980s

Although both light and electron microscopy continued to improve, it wasn't until the 1960s that researchers tried to combine the two imaging techniques. When searching the early literature for correlative microscopy publications, it becomes obvious that the term as we now use it, to indicate light and electron microscopic studies on the *same* area of the *same* sample, has evolved over time. The earliest references are frequently studies of the same tissue or sample type but not necessarily on the same specimen; thus, they are more comparative than truly correlative. The earliest paper that we have found that imaged a sample in a light microscope with a similarly prepared sample in an electron microscope is from the pioneering work of Keith Porter, where chick embryonic fibroblasts were cultured on a formvar substrate, fixed and imaged (Figure 1.1) [15]. This was only done as a proof of principle for developing EM techniques, though, and no attempt was made to draw conclusions from any correlation. A correlative study from 1960 by Goodman and Morgan was performed on separate cell cultures and published as two papers, one for light [16] and one for transmission electron microscopy (TEM)[17].

Other correlative studies from 1969 [18] and 1970 [19, 20] used biopsy samples that had been divided and processed for either light or electron microscopy, and then extrapolated between the morphological findings in each. Additional studies of correlative microscopy went a step further and used the same sample but adjacent sections. In

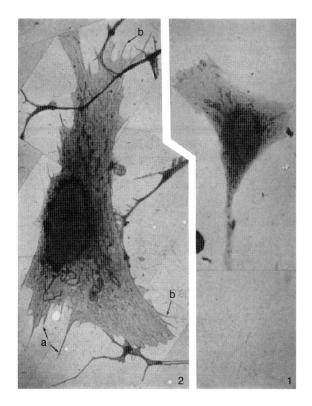

Figure 1.1 The first micrograph to compare a sample imaged with a light microscope; 1) and an electron microscope; 2), was published by Keith Porter in 194515. While not truly correlative, e.g. of the same specimen, this did demonstrate that samples prepared with the same procedures could be imaged using multiple methods. Reproduced with permission of ROCKEFELLER UNIVERSITY PRESS via Copyright Clearance Center ©1945.

1970, Watari and coworkers published a study of the islets of Langerhans using adjacent resin-embedded sections [21], and in 1979, Hyde et al. used the same block to first cut thick sections and inspect them by LM, then selected areas were cut out from these samples, and thin sectioned for TEM [22]. A very early attempt to combine immunohistochemistry with TEM was published in 1974 by Bordi and Bussolati [23].

In 1980, Gonda and Hsu combined LM, scanning electron microscopy (SEM) and TEM to study developing mouse blastocysts [24]. These early studies, although not meeting the criteria for correlative microscopy that we use today, were examples of researchers trying to use multiple microscopy methods to bridge the resolution gap between photons and electrons.

It was probably the 1967 article by McDonald, Pease, and Hayes [25] that examined sectioned rabbit tissues by LM and SEM, that marks the first use of correlative microscopy with the specific purpose of adding the extra resolution available in the EM to the LM data (Figure 1.2). A 1969 paper by McDonald and Hayes used fixed, dried blood cells and clots and correlated images of the same cells using their morphology and proximity to neighboring cells to identify them [26]. In 1971, a short technical note was published by Ayres, Allen, and Williams using specimens from a cervical biopsy that were inspected by SEM, then reprocessed for LM imaging [27]. At the same time, a group led by H.D. Geissinger at the University of Guelph was working intensely on correlative microscopy and matching the same area in cell preparations and tissue slices using SEM and LM [28]. In a 1973 paper, Geissinger, Basrur, and Yamashiro constructed a custom-built holder with a measuring caliper that could transfer between the LM stage and SEM specimen chamber [29,30], and by use of a method of correlated integers [31] they were able to reacquire the same coordinates and image the same area. Geissinger, Abandowitz, and Josefowicz used the same technique to examine hair shafts in transmitted and reflected light and SEM [32].

Geissenger and coworkers continued to explore the possibilities of correlative microscopy, combining many imaging modalities: SEM-interferometry [33], and LM polarization-SEM-TEM (Figure 1.3) [34]. This approach of combining multiple LM and EM techniques was also adopted by other investigators. A paper from 1989 used a combination of live cell video microscopy, low-voltage SEM, and high-voltage TEM to study membrane associated glycoproteins in human platelet cells [35], and that same year a paper used intravital video microscopy, LM and TEM to study capillary growth [36].

These early efforts to bridge the resolution scale were pioneering and led to a greater interest in using combined microscopic techniques to increase the data content of bioimaging experiments. This interest was further demonstrated in 1987 with the publication of the first book describing CLEM instrumentation and methods [37].

1.4 CLEM Matures as a Scientific Tool 1990 to 2017

In the next 25 years, the technique would continue to progress, not only with specific CLEM developments aimed at more precise and faster correlation but also by constantly implementing improvements made in the LM and EM fields. For example, in the early 1980s LM in the life sciences was reinvigorated by the development and commercialization of confocal laser scanning microscopy [38, 39]. This was quickly supplemented by the discovery of green fluorescent protein (GFP) [40, 41] and other

(a)

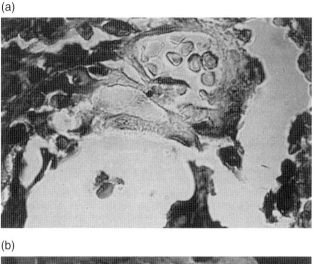

(b)

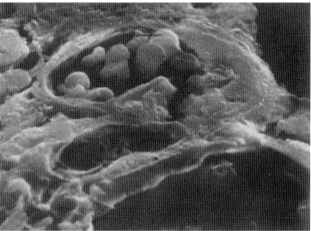

Figure 1.2 Quite possibly the first truly correlative micrograph, i.e, of the same sample imaged in both the LM (a) and EM (b), published by McDonald, Pierce and Hayes in 1967 [25]. An "area of delayed radiation lesion of rabbit sensory cortex" in a 4 μm section of paraffin embedded tissue. The authors used the correlation to point out features in the EM that were not obvious in the LM. Reprinted by permission of the publisher, Springer Nature, from laboratory Investigation © 1967.

fluorophores, allowing functional imaging in live cells [42, 43]. In the 1990s, subdiffrac-tion-limited or super-resolution LM techniques with resolutions below the limit set by Abbe's law started to appear. First was stimulated emission depletion (STED) micros-copy [44], followed by structured illumination microscopy (SIM) [45], stochastic optical reconstruction microscopy (STORM) [46], and photoactivated localization microscopy (PALM) [47] More recent developments such as light sheet microscopy [48], tissue clearing, expansion microscopy [49], and adaptive optics [50] continue to extend the capabilities of LM and in due course will find their way into CLEM.

On the EM side, electron tomography (ET), which offers 3D visualization of a selected part of a specimen at very high resolution, has been continuously improved since 1968

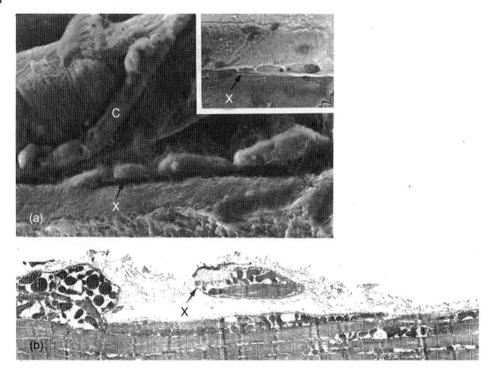

Figure 1.3 The laboratory of Professor H.D. Geissigner pioneered the development of different correlative microscopy workflows using multiple imaging modalities. This elegant example from a 1980 paper demonstrates the extra information to be gained through the correlation of LM (inset a × 400), SEM (a × 1500) and TEM (b × 5400) examining a sample of human muscle in a patient with muscular distrophy. Reprinted by permission of the publisher from Ultrastructural Pathology © 1980 (Taylor & Francis Ltd, http://www.tandfonline.com).

and with the growing computational power is still increasingly applied and optimized. One of the most recent developments is ET done under cryo-conditions [51, 52], providing exciting new applications in the field of structural biology. It is also of great importance for CLEM that the resolution of SEMs has been steadily improved, now almost approaching the TEM level [53]. Moreover, SEMs gained extended automated capabilities for 3D imaging [54, 55, 56], which greatly facilitates CLEM in the z-axis. Further improvements in LM and EM have made them more complementary in terms of resolution, contrast generation, and image dimensions. The promising power of combining these complementary modalities in integrated or modular microscopy settings has boosted the development and applications of CLEM in the last decade.

One of the main challenges in CLEM studies is to use LM to determine where you are in the landscape of the nanoworld of EM. Sadly for microscopists, a cellular version of GPS does not yet exist. To correlate a region of interest (ROI) in a light micrograph to the corresponding area in an electron micrograph is no easy task. Ideally, what is needed is a probe that can be easily visualized in both microscopes. An early but noncorrelative immunohistochemical study did develop such a probe using horseradish peroxidase (HRP)-antiperoxidase-diaminobenzidine (DAB), but only inspected it in the TEM [57].

To the best of our knowledge, the first demonstration of a directly correlative tracer, visible in both LM and EM, was in 1980 when Roth synthesized a FITC-protein A-colloidal gold complex to label antibodies to amylase in sections of pancreas [58]. In 1982, Maranto used photoconversion of the fluorescent dye Lucifer yellow in the presence of DAB to create an osmiophilic polymer [59], and this was followed by other studies using photooxidized fluorophores [60, 61]. In 1987, Quattrochi and colleagues synthesized fluorescent nanospheres linked by IgG to protein A-colloidal gold, and used them for retrograde neuronal labeling [62]. Polishchuk in 2000 used live-cell confocal microscopy to localize GFP to the Golgi complex and subsequent HRP-DAB reaction to relocate the ROI by serial section TEM to create 3D reconstructions [63]. In 2001, Adams reported the engineering of ReAsh-EDT$_2$ [64] a biarsenical ligand that could photooxidize tetracystine tagged GFP for use in CLEM studies [65]. In 2005, Grabenbauer demonstrated a new method that proved it was possible to use GFP to directly photooxidize DAB, thus allowing for the correlative visualization of endogenously expressed proteins [66]. This was followed in 2011 by the engineering of Mini-SOG, a fluorescent flavoprotein that was both fluorescent and a high-efficiency photooxidizer developed specifically by Roger Tsien for CLEM [67]. All these attempts were hampered by the diffusible, nonquantitative HRP DAB reaction product that decreased the precision of the correlation.

In 2012, Martell and co-workers developed APEX, a small (28-kDa) genetically engineered peroxidase that can be coupled to fluoroproteins, remains active following fixation, creates a more precise localized reaction and does not require light to reduce DAB [68]. APEX is now also available in a modular form incorporating a GFP binding protein [69]. In 2015, the laboratory of Ben Giepmans developed a probe called FLIPPER, expressly for CLEM studies containing a fluoroprotein and a peroxidase that could be genetically expressed [70]. Most recently Arnold et al. developed a cryo-LM stage that, combined with fiducial markers and a computational algorithm, is able to allow for precise correlation between cryo-LM and FIB-SEM [71] (Figure 1.4). This development is very exciting, as CLEM can now be used to precisely guide the FIB milling process of vitrified cellular samples and capture specific structures in their native orientation.

Since the interest in CLEM continues to increase, other types of probes are becoming rapidly available. In parallel and addition to the DAB-dependent probes already described, others tried to preserve fluorescent signals in resin-embedded tissues to make the CLEM workflow more flexible. In 2014, Peddie et al. succeeded in retaining fluoroprotein signals in resin embedded heavy metal stained tissues [72], and in the same year, Perkovic and co-authors published a similar method for organic fluorophores [73]. Super-resolution probes for CLEM have also been developed. In 2015, Paez Sengla and colleagues developed fixation-resistant photoactivatable fluoroproteins for CLEM [74]. Recently, scientists working in Jena, Germany, synthesized polylactide nanoparticles incorporating iridium (III) complexes [75], Müller et al. reported the use of self-labeling protein tags for time resolved CLEM experiments [76] and the laboratory of Roger Tsien described the use of Click-EM for imaging metabolically tagged non-protein biomolecules [77]. Nonfluorescent detection techniques to relocate ROIs have also been investigated. Physical marking of tissues using an infrared laser can help to reacquire an ROI at the EM level [78]. In 2012, Glenn and co-authors developed probes that are cathodoluminescent, raising the possibility of discriminating between multiple probes at the EM level [79]. In 2015, Nagayama et al. demonstrated eGFP cathodoluminescence [80], and

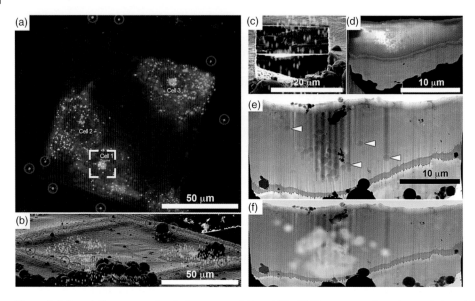

Figure 1.4 Correlation approaches between cryo-LM and FIB-SEM is very promising to precisely guide the FIB milling process of vitrified biological samples, and paves the way for visualization of single molecules in their native *in-situ* states. Arnold et al. developed a cryo stage for spinning-disk microscopy at cryogenic temperatures, and employed this approach to guide the FIB milling process to capture fluorescently labeled lipid droplets, in 300 nm lamellas. Reprinted by permission from: Arnold, Jan, et al. "Site-specific cryo-focused ion beam sample preparation guided by 3D correlative microscopy." *Biophysical Journal* 110.4; 2016): 860–869.

Furukawa and co-workers developed rare-earth nanophosphors for cathodoluminescent imaging in scanning-transmission EM [81]. In 2016, Fukushima and co-workers reported the development of yttrium oxide nanophosphors, which are both fluorescent and cathodoluminescent as well as electron dense [82] and Hemelaar et al. reported the use of fluorescent/cathodoluminescent nanodiamonds for correlative studies [83].

Although we still lack a cellular equivalent of GPS, a 2017 publication, integrating several of the developments mentioned above, has demonstrated a correlation technique using cathodoluminescence with a precision of <5 nm between LM and EM images [84]. While the development of probes is invaluable for the progress of CLEM, their applications are maximized thanks to the many computer vision and bioinformatics specialists who have made strides in the "back end" of the process; that of data analysis and eventual overlaying of the LM and EM digital data [85, 86, 87, 88]. With the development of specifically engineered probes and software, the CLEM workflow has become easier and more accessible to nonspecialist labs.

A way to minimize the effort to find back ROIs by LM and EM is to combine LM and EM in one microscope. The first attempt at such an integrated device was in 1978, when Hartmann and co-workers published a note describing an attachment to a commercial SEM that incorporated light microscope optics [89]. Further adaptations were made, and in 1981 JS Ploem presented a prototype Leitz instrument at the VIII Conference on Analytical Cytology and Cytometry [90]. In 1982, Wouters and Koerten also published an integrated instrument for LM and SEM [91]. Then, for over 20 years there was little progress on the instrumentation front, until in 2008 three groups from Utrecht

University and Leiden University Medical Center collaborated on the design of an adaptation to a TEM that incorporated a tilting sample holder and an aspherical (NA 0.55) objective lens mounted inside a TEM column perpendicular to the electron beam and connected externally to a scanning confocal microscope [92]. This integrated light and TEM was commercialized by FEI (now Thermo Fischer) under the name iCorr.

Although this approach worked successfully, it required special specimen preparation that had to compromise between preserving the fluorescent signal and maintaining ultrastructural integrity and contrast, which for room-temperature CLEM still is a challenge. An improvement on this design introducing cryo-EM to observe fluorochromes in their hydrated state was published in 2012 [93].

The following year a collaborative effort between groups from both academia and industry published the design for an integrated room-temperature fluorescence-SEM. This system employed a reflective mirror and a 45X NA 0.41 objective lens placed in an SEM column in which holes were bored to allow passage of the electron beam [94]. This design also required special specimen preparation with fixation in low-concentration gluteraldehyde and dehydration insensitive fluorochromes. As far as can be determined, this design was never commercialized.

In 2010, Nishiyama and co-workers published the design for a combined instrument in which a fluorescence microscope was mounted over an inverted SEM with the sample in a chamber constructed of a silicon nitride film [95]. This was later marketed by JEOL as the ClairScope. Because of its limited magnification and resolution capacity, it was never widely adopted but proved the impetus for other designs of integrated instruments.

In 2011, Albert Polman in AMOLF, Amsterdam developed the SPARC, a custom-built stage that could be installed in an SEM containing a swinging parabolic mirror connected to an external CCD camera recording cathodoluminescence [96]. This design was used primarily for materials science and nanophotonics applications, although recent breakthroughs in biological applications could change that [97]. In 2013, the Charged Particle Optics group of Delft University of Technology developed SECOM, which included a high NA optical lens inside the vacuum chamber directly below the SEM pole piece [98, 99]. Both SPARC and SECOM are marketed by a spinoff company, DELMIC BV. In 2013, DELMIC joined with Phenom-World BV to develop the Delphi that used a similar design to SECOM but as a standalone instrument; it was unveiled in 2014 at the IMC in Prague. A SECOM variant including super-resolution capability was released in 2016. Using this SECOM platform, Peddie et al. recently presented strong stable blinking properties of GFP and YFP *in-vacuo*, and used this to achieve super resolution CLEM of resin-embedded cells [100] (Figure 1.5). Additional developmental work is underway on integrated instrumentation such as the inclusion of a miniature fluorescence microscope into a serial blockface sectioning SEM system for 3D CLEM studies [101]; and with the growing interest in correlative microscopy we can expect to see new instruments with additional capabilities that could bring CLEM into more widespread use.

The early trend to combine multiple modes of light and electron microscopy continued into the 1990s with various permutations of light, fluorescence, video, SEM, and TEM being employed [102], and increasing numbers of studies that went beyond mere proof of principle to address significant scientific questions [103, 104, 105]. Newly developing microscopic techniques were integrated into CLEM studies including: intravital

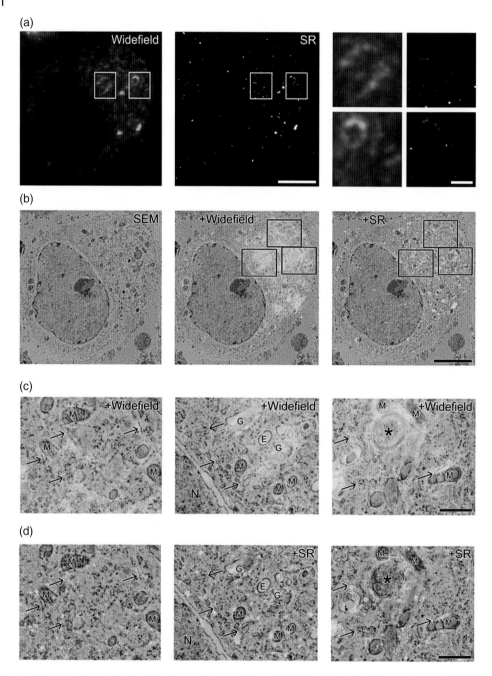

Figure 1.5 Correlative single molecule localization microscopy and EM of resin embedded cells in an integrated super-resolution LM in SEM. This approach successfully localizes the lipid DAG to subdomains of Golgi membranes, endoplasmic reticulum, cristae and outer mitochondrial membrane (c and d). Reprinted from Peddie, Christopher J, et al. "Correlative super-resolution fluorescence and electron microscopy using conventional fluorescent proteins *in vacuo*." *Journal of Structural Biology* 199.2: 120–131; 2017) under creative commons license 4.0.

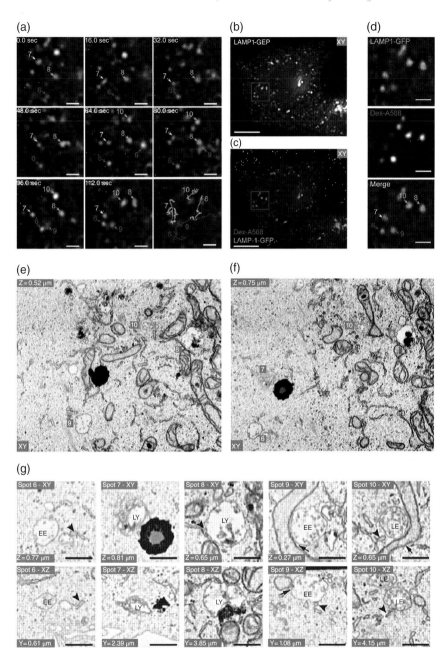

Figure 1.6 Correlation of live-cell imaging to 3D-EM integrates multiple dynamic and ultrastructural parameters at the cellular and subcellular level. The Klumperman laboratory recently published the first example of CLEM from live cells to 3D EM at the level of a single organelle. The 3D EM data (e–g) are obtained by FIB.SEM of the ROI identified in live cells (a–c) and fluorescence microscopy (D). Reprinted by permission from: Fermie, et al. "Single organelle dynamics linked to 3D structure by correlative live-cell imaging and 3D electron microscopy." *Traffic* 19.5; 2018): 354–369.

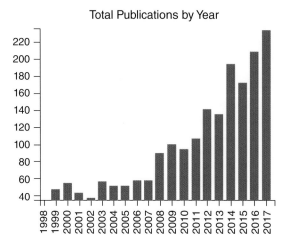

Total Publications by Year

Figure 1.7 Citation report generated from Web of Science Core Collection between years 1900–2017 for 4,139 publications, which have the words *correlative, light, electron,* and *microscopy* either in their title or topic.

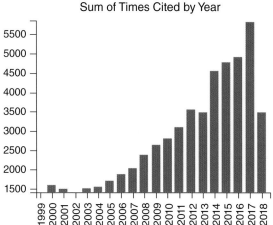

Sum of Times Cited by Year

two-photon [106], florescence recovery after photobleaching (FRAP) [107], cryo-LM, and cryo-EM [108], live-cell imaging and immunogold labeling [109], fluorescence in situ hybridization (FISH) [110], cryo-scanning transmission electron microscopy (cryo-STEM) [111], and most recently super-resolution microscopy [112, 113, 114, 115, 116, 117, 118]. For a long time, CLEM had remained a 2D technique, at least at the EM end. With the advent of TEM tomography, this changed, and many correlative LM-TEM tomographic studies have opened up the third dimension at the nanoscale [119, 120, 121, 122, 123, 124, 125, 126]. More recently, other 3D EM techniques have been used for CLEM, such as: serial section TEM (ssTEM) [127, 128], array tomography [129, 130], focused ion beam SEM (FIB-SEM) [131, 132, 133, 134], and serial blockface imaging (SBFI) [135]. Fermie et al. demonstrated correlation of live-cell imaging to 3D EM, and linked dynamic characteristics of single organelles to their 3D ultrastructure [136](Figure 1.6). Other imaging techniques have also been used to enhance LM and EM studies, most often the use of CT or soft X-rays produced by either synchrotron beamlines or standalone instrumentation [137, 138, 139, 140].

CLEM came about because scientists needed more information than either LM or EM could offer on its own. The technique began slowly and it was some time before the hardware and software available could fully support its development. What began as a few papers showcasing the significant efforts of some intrepid microscopists has now become a rapidly growing body of literature (Figure 1.7). It took many years for photons and electrons to learn to work together, now that they have, the future for correlative microscopy looks bright indeed.

Acknowledgments

The authors would like to thank Emeritus Professor Shigeto Yamashiro, Department of Biomedical Sciences, Ontario Veterinary College, University of Guelph, and Dr. Henry Abandowitz for sharing their recollections of the early years of CLEM research and of the late Professor H. Dieter Geissenger.

References

1 Lindberg, DC. Theories of Vision from Al-Kindi to Kepler. Chicago: University of Chicago Press; 1976.
2 The Life and Work of Athanaseus Kircher, S.J. http://mjt.org/exhibits/kircher.html.
3 van Zuylen J. The microscopes of Antoni Van Leeuwenhoek. J Microsc. 1981 Mar;121(Pt 3):309–28.
4 Tierie G. Corneilus Drebbel; 1572–1633. H.J. Paris; 1932. ASIN: B00086L36G.
5 Court TH, and Clay RS. The history of the microscope. London: Charles Griffin and Company; 1932.
6 Hooke R. London: Microscopia; 1665.
7 Abbe E. On new methods for improving spherical correction applied to the construction of wide-angled object-glasses. Journal of the Royal Microscopical Society 1879 2(7):812–24.
8 Abbe E. "Über einen neuen Beleuchtungsapparat am Mikroskop" [About a New Illumination Apparatus to the Microscope]. Archiv für mikroskopische Anatomie (in German). Bonn, Germany: Verlag von Max Cohen & Sohn; 1873 9:469–80.
9 Köhler A. "Ein neues Beleuchtungsverfahren für mikrophotographische Zwecke". Zeitschrift für wissenschaftliche Mikroskopie und für Mikroskopische Technik. 1893 10(4):433–40.
10 Abbe E. "Beiträge zur Theorie des Mikroskops und der mikroskopischen Wahrnehmung" [Contributions to the Theory of the Microscope and of Microscopic Perception]. Archiv für Mikroskopische Anatomie. Bonn, Germany: Verlag von Max Cohen & Sohn. 1873 9(1):413–68.
11 De Brogle L. Recherches sur la Théorie des Quanta. Ph.D. Thesis, Paris University; 1924.
12 Van Dyck D. Electron microscopy in Belgium. Advances in Imaging and Electronic Physics 1996 96:67–78.
13 Ruska E. Nobel lecture. The development of the electron microscope and of electron microscopy. Biosci Rep. 1987 Aug;7(8):607–29.
14 Marton L. Electron Microscopy of biological objects. Nature 1934 133, 911.
15 Porter KR, Claude A, and Fullam EF. A study of tissue culture cells by electron microscopy. J. Exp. Med. 1945 81(3):233–46.

16 Goodman GC, Morgan C, Breitenfeld PM, and Rose HM. A correlative study by electron and light microscopy of the development of the type 5 adenovirus. II Light microscopy J. Exp. Med. 1960 112:383–402.

17 Morgan C, Goodman GC, Breitenfeld PM, and Rose HM. A correlative study by electron and light microscopy of the development of the type 5 adenovirus. I Electron microscopy J. Exp. Med. 1960 112:373–82.

18 Black WC. A Correlative light and electron microscopy in primary hyperparathyroidism. Arch. Path. 1969 88:225–41.

19 Fish AJ, Herdman RC, Michael AF, Pickering RJ, and Good RA. Epidemic acute glomerulonephritis associated with type 49 streptococcal pyroderma. II Correlative study of light, imunofluorescent and electron microscopic findings. Am. J. Med. 1970 48:28–39.

20 Burkholder PM, and Bergeron JA. Spontaneous Glomerulonephritis in the prosimian primate Galago. Am. J. Path. 1970 61(3):450.

21 Watari N, Tsukagoshi N, and Honma Y. The correlative light and electron microscopy of the islets of Langerhans in some lower vertebrates. Arch. Histol. Jap. 1970 31 (3/4):371–92.

22 Hyde DM, Samuelson A, Blakeney WH, and Kosch PC. A correlative light microscopy, transmission and scanning electron microscopy study of the ferret lung. Scanning Electron Microscopy 1979 3:891–98.

23 Bordi C, and Bussolati G. Immunofluorescent, histochemical and ultrastructural studies for the detection of multiple endocrine polypeptide tumors of the pancreas. Virchows. Arch. B Cell Path. 1974 17:13–27.

24 Gonda MA, and Hsu Y. Correlative scanning electron, transmission electron and light microscopic studies of the in vitro development of mouse embryos on a plastic substrate at the implantation stage. J. Embryol. Exp. Morph. 1980 56:23–39.

25 McDonald LW, Pease RFW, and Hayes TL. Scanning electron microscopy of sectioned tissue. Laboratory Invest. 1967 16(4):532–38.

26 McDonald LW, and Hayes TL. Correlation of scanning electron microscope and light microscope images of individual cells in human blood and blood clots. Exp and Molecular Path. 1969 10:186–98.

27 Ayres A, Allen JM, and Williams AE. A method for obtaining conventional histological sections from specimens after examination by scanning electron microscopy. J. Micros. 1971 93(3):247–50.

28 Geissinger HD, and Bond EF. Nomarski differential interference contrast microscopy and scanning electron microscopy of tissue sections and fibroblast cell culture monolayers. Mikroskopie 1971 27:32–39.

29 Geissinger HD, Basrur PK, and Yamashiro S. Fast scanning electron microscopic and light microscopic correlation of paraffin sections and chromosome spreads of nickel induced tumor. Trans. Am. Micros. Soc. 1973 92(2):209–17.

30 Geissenger HD. A precise stage arrangement for correlative microscopy for specimens mounted on glass slides stubs or EM grids. J. Microsc. 1974 100:113–17.

31 Geissinger HD, and Kamler J. Precise and fast correlation of light microscopic and scanning electron microscopic images. Can. Res. Dev. 1972 5:13–16.

32 Geissinger HD, Abandowitz HM, and Josefowicz WJ. Correlative light optical and scanning electron microscopy of single hair strands. Mikroskopie 1975 31: 279–286.

33 Abandowitz HM, and Geissenger HD, Preparation of cells from suspensions for correlative scenning electron and interference microscopy. Histochemistry; 1975 45:89–94.

34 Geissenger HD, Vriend RA, Ackerley CA, and Yamashiro S. Correlative light optical, scanning electron and transmission electron microscopy of skeletal muscle in muscular dystrophy and muscular atrophy: A pilot study. Ultrastructural Pathology 1980 1:327–35.

35 Albrecht RM, Goodman SL, and Simmons SR. Distribution and movement of membrane-associated platelet glycoproteins: Use of colloidal gold with correlative video enhanced light microscopy low-voltage high-resolution scanning electron microscopy and high-voltage transmission electron microscopy. Am. J. Anat. 1989 185:149–64.

36 Rhodin JAG, and Fujita H. Capillary growth in the mesentery of normal young rats. Intravital video and electron microscopeanalysis. J submicrosc. Cytol. Pathol. 1989 21(1):1–34.

37 Correlative microscopy in biology. Instrumentaiton and methods. Hyatt MA ed. Academic Press Inc. Orlando, Florida; 1987.

38 Minsky M. Memoir on inventing the confocal scanning microscope. Scanning 1988 10:128–138.

39 Sheppard CJR, and Wilson T. The theory of the direct-view confocal microscope. J. Microsc. 1981 124(2):107–17.

40 Morise H, Shimomura S, Johnson FH, and Winant J. Imtermolecular energy transfer in the bioluminescent system of Aequorea. Biochemistry 1974 13(12):2656–62.

41 Chalfie M, Tu Y, Euskirchen G, Ward WW, Prasher DC, Green fluorescent protein as a marker for gene expression. Science 1994 263(5148):802–05.

42 Grynkiewicz G, Poenie M, Tsien RY. A new generation of Ca2+ indicators with greatly improved fluorescence properties. J. Biol. Chem. 1985 260(6):3440–50.

43 Giepmans BNG, Adams SR, Ellisman MH, and Tsien RY. The fluorescent toolbox for assessing protein location and function. Science 2006:217–24.

44 Hell SW, Wichmann J. Breaking the diffraction resolution limit by stimulated emission: stimulated-emission-depletion fluorescence microscopy. Opt. Lett. 1994 19(11),780–82.

45 Gustafsson MGL. Nonlinear structured-illumination microscopy: wide-field fluorescence imaging with theoretically unlimited resolution. Proc. Natl. Acad. Sci. 2005 102(37):1308186.

46 Rust M, Bates M, and Zhuang X. Sub-diffraction-limit imaging by stochastic optical reconstruction microscopy (STORM). Nature Methods. 2006 3(10):793–96.

47 Betzig, E, Patterson GH, Sougrat R, Wolf Lindwasser O, Olenych S, Bonifacino JS, et al. Imaging Intracellular Fluorescent Proteins at Nanometer Resolution. Science. 2006 313 (5793):1642–45.

48 Voie AH, Burns DH, and Spelman FA. Orthogonal plane fluorescence optical sectioning: three-dimensional imaging of macroscopic biological specimens. J. Microsc. 1993 170(3):229–36.

49 Marx V. Optimizing probes to image cleared tissue. Nat. Meth. 2016 13(3):205–09.

50 Wang K, Milkie DE, Saxena A, Engerer P, Misgeld T, Bronner M, et al. Rapid adaptive optical recovery of optimal resolution over large volumes. Nat. Meth. 2014 11:625–28.

51 Weyland M, and Midgley PA. Electron Tomography. Material Today 2004 7(12):32–40.

52 Al-Amoudi A, Chang J-J, Leforestier A, McDowall A, Salamin LM, Norlén LPO, et al. Cryo-electron Microscopy of Vitreous Sections. EMBO J. 2004 23(18):3583–88.

53 Rice WL, Van Hoek AN, Păunescu TG, Huynh C, Goetze G, Singh B, et al. High-resolution helium ion scanning microscopy of the rat kidney. PLOS ONE 2013 8(3):e57051.

54 Leighton SB. SEM images of block faces, cut by a miniature microtome within the SEM—A technical note. Scan Electron Microsc. 1981 2:73–76.

55 Armer HE, Mariggi G, Png KM, Genoud C, Monteith AG, et al. Imaging transient blood vessel fusion events in zebrafish by correlative volume electron microscopy PLOS ONE 2009 4:e7716.

56 Heymann JA, Shi D, Kim S, Bliss D, Milne JL, and Subramaniam S. 3D imaging of mammalian cells with ion-abrasion scanning electron microscopy. J. Struct. Biol. 2009 166:1–7.

57 Moriarty GC, and Halmi NS. Electron microscopic study of the adrenocorticotropin-producing cell with the use of unlabeled antibody and the soluble peroxidase-antiperoxidase complex. J. Histochem. Cytochem. 1972 20(8): 590–603.

58 Roth J, Bendayan M, and Orci L. FITC-protein a-gold complex for light and electron microscopic immunocytochemistry. J Histochem. Cytochem. 1980 28:55–57.

59 Maranto AR. Neuronal Mapping: A photooxidation reaction makes Lucifer yellow useful for electron microscopy. Science 1982 217:953–55.

60 Sandell JH, and Masland RH. Photoconversion of some fluorescent markers to a diaminobenzadine product. J. histochem. Cytochem. 1988 36: 555–559.

61 Deerinck TJ, Martone ME, Lev-Ram M, Green DPL, Tsien RY, and Spector DL. Fluorescence photooxidation with eosin: A method for high resolution immunolocalization and in situ hybridization detection for light and electron microscopy. J. Cell Biol. 1994 126(4): 901–10.

62 Quattrochi JJ, Madison R, Sidman RL, and Kijavin I. Colloidal gold fluorescent microspheres: a new retrograde marker visualized by light and electron miceoscopy. Exp. Neurol. 1987 96:219–24.

63 Polishchuk RS, Polishchuk EV, Marra P, Alberti S, Buccione R, Luini A. et al. Correlative light-electron microscopy reveals the tubular-saccular ultrastructure of carriers operating between golgi apparatus and plasma membrane. J. Cell Biol. 2000 148(1):45–58.

64 Adams SR, Campbell RE, Gross LA, Martin BR, Walkup GK, Yao Y, et al. New biarsenical ligands and tetracystine motifs for protein labeling in vivo and in vitro: synthesis and biological applications. J. Am. Chem. Soc. 2001 124:6063–76.

65 Gaietta GM, Giepmans BNG, Deerinck TJ, Smith WB, Ngan L, Llopis J, et al. Golgi twins in late mitosis reveled by genetically encoded tags for live cell imaging and correlated electron microscopy. Proc. Nat. Acad. Sci. 2006 103(47):17777–82.

66 Grabenbauer M, Geerts WJC, Fernandez-Rodriguez J, Hoenger A, Koster A.J, and Nilsson T. Correlative microscopy and electron tomography of GFP through photooxidation. Nat. Meth. 2005 2(11):857–62.

67 Shu X, Lev-Ram V, Deerinck TJ, Oi Y, Ramko E.B, Davidson MW et al. A genetically encoded tag for correlated light and electron microscopy of intact cells, tissues, and organisms. PLOS Biology 2011 9(4):e100141.

68 Martell JD, Deerinck TJ, Sancak Y, Poulos TL, Mootha VK, Sosinsky GE, et al. Engineered ascorbate peroxidase as a genetically encoded reporter for electron microscopy. Nat. Biotechnol. 2012 30(11):1143–51.

69 Ariotti N, Hall TE, and Parton RG. Correlative light and electron microscopic detection of GFP-labeled proteins using modular APEX. Methods cell biol. 2017 140:105–121.

70 Kuipers J, van Ham TJ, Kalicharan RD, Veenstra-Algra A, Sjollema KA, Dijk F, Schnell U, et al. FLIPPER, a combinatorial probe for correlated live imaging and electron microscopy, allows identification and quantitative analysis of various cells and organelles. Cell Tiss. Res. 2015 360(1):61–70.

71 Arnold J, Mahamid J, Lucic V, de Marco A, Fernandez J-J, Laugks T, et al. Site-specific cryo-focused ion beam sample preparationguided by 3D correlative microscopy. Biophysical J. 2016 110:860–869.

72 Peddie CJ, Blight K, Wilson E, Melia C, Marrison J, Carzaniga R, et al. Correlative and integrated light and electron microscopy of in-resin GFP fluorescence, used to localize diacylglycerol in mammalian cells. Ultramicroscopy. 2014 143:3–14.

73 Perkovic M, Kunz M, Endesfelder U, Bunse S, Wigge C, Yu Z, et al. Correlative light-electron microscopy with chemical tags. J. Struct. Biol. 2014 186:205–13.

74 Paez Segala MG, Sun MG, Shtengel G, Viswanathan S, Baird MA, Macklin JJ, et al. Fixation-resistant photoactivatable fluorescent proteins for correlative light and electron microscopy. Nat. Methods. 2015 12(3):215–218.

75 Reifarth M, Pretzel D, Schubert S, Weber C, Heintzmann R, Hoeppener S, et al. Cellular uptake of PLA nanoparticles studied by light and electron microscopy: synthesis, characterization and biocompatibility studies using an iridium (III) complex as a correlative label. Chem. Commun. 2016 52:4361–4364.

76 Müller A, Neukam M, Ivanova A, Sönmez A, Münster C, Kretschmar S, et al. Super-resolution and electron microscopy on cryo and epoxy sections using self-labeling protein tags. Sci. Rep. 2017 Dec;7(1):23. doi: 10.1038/s41598-017-00033-x.

77 Ngo JT, Adams SR, Deerinck TJ, Boassa D, Rodriguez-Rivera F, Palida SF, et al. Click-electron microscopy for imaging metabolically tagged non-protein biomolecules. Nat. Chem. Biol. 2016 12(6) 459–65.

78 Bishop D, Nikić I, Brinkoetter M, Knecht S, Potz S, Kerschensteiner M, et al. Near-infrared branding efficiently correlates light and electron microscopy. Nat. Methods 2011 8(7):568–72.

79 Glenn DR, Zhang H, Kasthuri N, Schalek R, Lo PK, Trifonov AS, et al. Correlative light and electron microscopy using cathodoluminescence from nanoparticles with distinguishable colours. Sci. Reports 2012: 865 doi:10.1038/srep00865.

80 Nagayama K, Onuma T, Ueno R, Tamehiro K, Minoda H. Cathodoluminescence and electron-induced fluorescence enhancement of enhanced green fluorescent protein. J. Physical Chem. B. 2016 120:1169–74.

81 Furukawa T, Fukushima S, Niioka H, Yamamoto N, Miyake J, Araki T, et al. Rare-earth doped nanophosphors for multicolor cathodoluminescence nanobioimaging using scanning transmission electron microscopy. J. Biomed. 2015 Oct 20(5):doi: 10.1117/1. JBO.20.5.056007.

82 Fukushima S, Furukawa T, Niioka H, Ichimiya M, Sannomiya T, Tanaka N, et al. Correlative near-infared light and cathodoluminescence microscopy using Y_2O_3:Ln, Yb (Ln = Tm, Er) nanophosphors for multiscale multicolour bioimaging. Sci. Rep. 2016 6:25950.

83 Hemelaar SR, de Boer P, Chipaux M, Zuidema W, Hamoh T, Perona Martinez F, et al. Nanodiamonds as multi-purpose labels for microscopy. Scientific reports. 2017 Apr 7;7(1):720.

84 Haring MT, Liv N, Zonnevylle AC, Narvaez AC, Voortman LM, Kruit P, et al. Automated sub-5nm image registration in integrated correlative fluorescence and electron microscopy using cathodoluminescence pointers. Sci. Rep. 2017 7:43621 | DOI: 10.1038/srep43621

85 Kremer JR, Mastronarde DN, and McIntosh JR. Computer visualization of three-dimensional image data using IMOD. J. Structural Biol. 1996 116:71–76.

86 Keene DR, Tufa SF, Lunstrum GP, and Holden P. Confocal/TEM overlay microscopy: a simple method for correlating confocal and electron microscopy of cells expressing GFP/YFP fusion proteins. Microsc. Microanalys. 2008 14(4):342–48.

87 Cao T, Zach C, Molda S, Powell D, Czymmek K, and Neithammer M. Multi-modal registration for correlative microscopy using image analogies. Med. Image Analysis 2014 18:914–26.

88 Brama E, Peddie CJ, Wilkes G, Gu Y, Collinson LM, and Jones ML. ultraLM and miniLM: locator tools for smart tracking of fluorescent cells in correlative light and electron microscopy. Welcome Open Research 2016 1:26 doi: 10.12688/wellcomeopenres.10299.1.

89 Hartmann H, Hund A, Moll SH, and Thaer A. Attachment for combined scanning electron and light microscopical examinations. Beitr. Elektronenmikroskop. Direktabb. Oberfl. 1978 11:381–88.

90 Ploem JS. A new instrument permitting simultaneous scanning electron microscopy and fluorescence microscopy of the same specimen by integrating a LM in a SEM vacuum chamber. Cytometry 1981 2:121.

91 Wouters, C, and Koerten, H. Combined Light Microscope and Scanning Electron Microscope, A New Instrument for Cell Biology. Cell Biol. Int. Rep. 1982 6(10):955–59.

92 Agronskaia AV, Valentijn JA, van Driel LF, Schneijdenberg CTM, Humbel BM, van Bergen en Henegouwen PMP, et al. Integrated fluorescence and transmission electron microscopy. J. Structural Biol. 2008 164:183–89.

93 Faas FGA, Bárcena M, Agronskaia AV, Gerritsen H.C, Moscicka KB, Diebolder C.A, et al. Localization of fluorescently labeled structures in frozen-hydrated samples using integrated light electron microscopy. J Structural Biol. 2012 181:283–90.

94 Kanemaru T, Hirata K, Takasu S, Isobe S, Mizuki M, Mataka S. et al. A fluorescence scanning electron microscope. J. Structural Biol. 2009 109:344–49.

95 Nishiyama H, Suga M, Ogura T, Maruyama Y, Koizumi M, Mio K, et al. Atmospheric scanning electron microscope observes cells and tissues in open medium through silicon nitride film. J. Structural. Biol. 2010 169(3):438–49.

96 Coenen T, Vesseur EJR, and Polman A. Angle-resolved cathodoluminescence spectroscopy. Appl. Phys. Lett. 2011 99(14):143103.

97 Keevend K, Stiefel M, Neuer AL, Matter MT, Neels A, Bertazzo S, et al. Tb3+-doped LaF3 nanocrystals for correlative cathodoluminescence electron microscopy imaging with nanometric resolution in focused ion beam-sectioned biological samples. Nanoscale 2017 9:4383–87.

98 Zonnevylle AC, Van Tol RFC, Liv N, Narvarez AC, Effting APJ, Kruit P, et al. Integration of a high-NA light microscope in a scanning electron microscope. J. Micros. 2013 252(1):58–70.

99 Liv N, Zonnevylle C, Narvaez AC, Effting AP, Voorneveld PW, Lucas MS, et al. Simultaneous correlative scanning electron and high-NA fluorescence microscopy. PLOS One 2013 8(2):e55707.

100 Peddie CJ, Domart M-C, Snetkov X, O'Toole P, Larijani B, Way M, et al. Correlative super-resolution fluorescence and electron microscopy using conventional fluorescent proteins in vacuo J. Structural Biol. 2017 199:120–31.

101 Brama E, Peddie CJ, Wilkes G, Gu Y, Collinson L, and Jones ML. ultraLM and miniLM: Locator tools for smart tracking of fluorescent cells in correlative light and electron microscopy. Wellcome Open Research 2016 1:26.

102 Simmons SR, Pawley JB, and Albrecht RM. Optimizing parameters for correlative immunogold localization by video-enhanced light microscopy, high-voltage transmission electron microscopy and field emission scanning electron microscopy. J Histochem. Cytochem. 1990 38(12):1781–85.

103 Rieder CL, Cole RW, Khodjakov A, and Sluder G. The checkpoint delaying anaphase in response to chromosome monoorientation is mediated by an inhibitory signal produced by unattached kinetochores. J. Cell Biol. 1995 130(4):941–48.

104 Polishchuk RS, Polishchuk EV, and Minonov AA. Coalescence of Golgi fragments in microtubule-deprived living cells. Eur. J. Cell Biol. 1999 78:170–85.

105 Mironov AA, Polishchuk RS, and Luini A. Visualizing membrane traffic in vivo by combined video fluorescence and 3D electron microscopy. Trends Cell Biol. 2000 10:349–53.

106 Sandoval RM, Kennedy MD, Low PS, and Molitoris BA. Uptake and trafficking of fluorescent conjugates of folic acid in intact kidney determined using intravital two-photon microscopy. Am J. Cell Physiol. 2004 287:C517–26.

107 Darcy KJ, Staras K, Collinson LM, and Goda Y. An ultrastructural readout of fluorescence recovery after photobleaching using correlative light and electron microscopy. Nat. Protocols 2006 1(2):988–94.

108 Schwartz CL, Sarbash VI, Ataullakhanov FI, Mcintosh R, and Nicastro D. Cryo-fluorescence microscopy facilitates correlations between light and cryo-electron microscopy and reduces the rate of photobleaching. J. Microsc. 2007 227(2):98–109.

109 van Rijnsoever C, Oorschot V, and Klumperman J. Correlative light-electron microscopy (CLEM) combining live-cell imaging and Immunolabeling of ultrathin cryosections. Nat. Methods. 2008 5(11):973–80.

110 Halary S, Duperron S, and Boudier T. Direct image-based correlative microscopy technique for coupling identification and structural investigation of bacterial symbiosis associated with metazoans. Appl. And Environmental Micro. 2011 77(12):4172–79.

111 Nolin F, Ploton D, Wortham L, Tchelidze P, Bobichon H, Banchet V, et al. Targeted nano analysis of water and ions in the nucleus using cryo-correlative microscopy. Meth. Mol. Biol. 2015 1228:145–58.

112 Betzig E, Patterson GH, Sougrat R, Lindwasser OW, Olenych S, Bonifacino JS, et al. Imaging intracellular proteins at nanometer resolution. Science 2006 313:1642–45.

113 Perinetti G, Müller T, Spaar A, Polishchuk R, Luini A, and Egner A. Correlation of 4Pi and electron microscopy to study transport through single Golgi stacks in living cells with super resolution. Traffic. 2009 10:379–91.

114 Watanabe S, Richards J, Hollopeter G, Hobson RJ, Davis WM, and Jorgensen EM. Nano-fEM: protein localization using photo-activated localization microscopy and electron microscopy. JoVE 2012 Dec 3(70):e3995.

115 Kopek B, Shtengel G, Grimm JB, Clayton DA, and Hess HF. Correlated photoactivated localization and scanning electron microscopy. PLOS One 2013 8(10):e77209.

116 Ligeon L-A, Barois N, Werkmeister E, Bongiovanni A and Lafont F. Structured illumination microscopy and correlative microscopy to study autophagy. Methods 2015 75:61–68.

117 Shtengel G, Wang Y, Zhang Z, Goh W, Hess HF, and Kanchanawong P. Imaging cellular ultrastructure by PALM, iPALM and correlative iPALM-EM. Meth. Cell Biol. 2014 123:273–94.

118 Kim D, Deerinck TJ, Sigal YM, Babcock HP, Ellisman MH, and Zhuang X. Correlated stochastic optical reconstruction microscopy and electron microscopy. PLOS ONE 2015 10(4):e0124581. doi: 10.1371/journal.pone.0124581.

119 Sartori A, Gatz R, Beck F, Rigort A, Baumeister W, and Plitzko JM. Correlative microscopy: bridging the gap between fluorescence light microscopy and cryo-electron tomography. J. Structural Biol. 2007 160:135–45.

120 Lučić V, Kossel AH, Yang T, Bonhoeffer T, Baumeister W, and Sartori A. Multiscale imaging of neurons grown in culture: From light microscopy to cryo-electron tomography. J. Structural Biol. 2007 160:146–56.

121 Vicidomini G, Gagliani MC, Canfora M, Cortese K, Frosi F, Santangelo C, et al. High data output and automated 3D correlative light-electron microscopy method. Traffic. 2008 (9):1828–38.

122 van Driel LF, Valentijn JA, Valentijn KM, Koning RI, and Koster AJ. Tools for correlative cryo-fluorescence microscopy and cryo-electron tomography applied to whole mitochondria in human endothelial cells. Eur. J. Cell Biol. 2009 88:669–84.

123 Nixon SJ, Webb RI, Floetenmeyer M, Schieber N, Lo HP, and Parton RG. A single method for cryofixation and correlative light, electron microscopy and tomography of Zebrafish embryos. Traffic 2009 10:131–36.

124 Spiegelhalter C, Tosch V, Hentsch D, Koch M, Kessler P, Schwab Y, et al. From dynamic live cell imaging to 3D ultrastructure: Novel integrated methods for high pressure freezing and correlative light-electron microscopy. PLOS ONE 2010 5(2):e9014.

125 Liu B, Xue Y, Zhao W, Chen Y, Fan C, Gu L, et al. Three-dimensional super-resolution protein localization correlated with vitrified cellular context. Sci Rep. 2015 5:13017 | DOI: 10.1038/srep13017.

126 Schorb M, Gaechter L, Avinoam O, Sieckmann F, Clarke M, Bebeacua C, et al. New hardware and workflows for semi-automated correlative cryo-fluorescence and cryo-electron microscopy/tomography. J. Structural Biol. 2017 197:83–93.

127 Knott GW, Holtmaat A, Wilbrecht L, Welker E, and Svoboda K. Spine growth precedes synapse formation in the adult neocortex in vivo. Nat. Neurosci. 2006 9(9):1117–24.

128 Karreman MA, Mercier L, Schieber NL, Shibue T, Schwab Y, and Goetz JG. Correlating intravital multi-photon microscopy to 3D electron microscopy of invading tumor cells using anatomical reference points. PLOS ONE 2014 DOI: 10.1371/journal.pone.0114448.

129 Micheva KD, and Smith SJ. Array Tomography: A new tool for imaging the molecular architecture and ultrastructure of neural circuits. Neuron 2007 55(1):25–36.

130 Markert SM, Britz S, Proppert S, Lang M, Witvliet D, Mulcahy B, et al. Filling the gap: adding super-resolution to array tomography for correlated ultrastructural and molecular identification of electrical synapses at the C. elegans connectome. Neurophoton. 2016 3(4):041802 doi: 10.1117/1.NPh.3.4.041802.

131 Murphy GE, Narayan K, Lowekamp BC, Hartnell LM, Heymann JAW, Fu J, et al. Correlative 3D imaging of whole mammalian cells with light and electron microscopy. J. Struct. Biol. 2011 176:268–78.

132 Bushby AJ, Mariggi G, Armer HEJ, and Collinson LM. Correlative light and volume electron microscopy: using focused ion beam scanning electron microscopy to image transient events in model organisms. Meth. Mol. Biol. 2012 111:357–82.

133 Llorca L.B, Hummel E, Zimmerman H, Zou C, Burgold S, Reitdorf J, and Herms J. Correlation of two photon in vivo imaging and FIB/SEM microscopy. J Microsc. 2015 259(2):129–36.

134 Keevend K, Stiefel M, Neuer AL, Matter MT, Neels A, Bertazzo S, and Herrmann I.K, Tb3+doped LaF$_3$ nanocrystals for correlative cathodoluminescence electron microscopy imaging with nanometric resolution in focused ion beam-sectioned biological samples. Nanoscale, 9:4383–4387; 2017)

135 Urwyler O, Izadifar A, Dascenco D, Petrovic M, He H, Ayaz D, et al. Investigating CNS synaptogenesis at single-synapse resolution by combining reverse genetics with correlative light and electron microscopy. Development 2015 142:394–405.

136 Fermie J, Liv N, ten Brink C, van Donselaar E, Müller WH, Schieber N, et al. Single organelle dynamics linked to 3D structure by correlative live-cell imaging and 3D electron microscopy. Traffic 2018 19:354–69.

137 Hagen C, Guttmann P, Klupp B, Werner S, Rehbein S, Mettenleiter TC, et al. Correlative Vis-fluorescence and soft X-ray cryo-microscopy/tomography of adherent cells. J. Structural Biol. 2012 177:193–201.

138 Sengle G, Tufa SF, Sakai LY, Zulliger MA, and Keene DR. A correlative method for imaging identical regions of samples by micro-CT, light microscopy, and electron microscopy: imaging adipose tissue in a model system. Journal of Histochemistry & Cytochemistry. 2013 Apr;61(4):263–71.

139 Bushong EA, Johnson DD, Kim K-Y, Terada M, Hatori M, Peltier ST, et al. X-ray microscopy as an approach to increasing accuracy and efficiency of serial block-face imaging for correlated light and electron microscopy of biological specimens. Microsc. Microanal. 2015 21(1):231–38.

140 Sorrentino A, Nicolás J, Valcárcel R, Chichon FJ, Rosanes M, Avila J, et al. MISTRAL: a transmission soft X-ray microscopy beamline for cryo nano-tomography of biological samples and magnetic domains imaging. J. Synchrotron radiation 2015 22:1112–17.

2

Challenges for CLEM from a Light Microscopy Perspective

Kurt Anderson[1], Tommy Nilsson[2], and Julia Fernandez-Rodriguez[3]

[1] *The Francis Crick Institute, London, United Kingdom*
[2] *The Research Institute of the McGill University Health Centre and McGill University, Montreal, Canada*
[3] *Centre for Cellular Imaging at Sahlgrenska Academy, University of Gothenburg, Sweden*

2.1 Introduction

2.1.1 Electron and Light Microscopy

The eukaryotic cell relies on a highly regulated and functionally distinct set of membrane bound organelles that serve to compartmentalize reactions and preserve the biochemical polarity necessary for proper cellular function. When elucidating molecular function, important information can be derived from observing where this takes place, at the subcellular level. Indeed, an overwhelming amount of key knowledge has been derived by correlating cellular structures to function at the ultrastructural level using electron microscopy (EM), and this dates back to the 1950s. Of major importance, EM has enabled the visualization of the ultrastructural context of protein synthesis, vesicular transport, and processing through the secretory pathway as well as endocytosis and protein degradation in lysosomes. Early three-dimensional (3D) reconstructions also provided beautiful and intricate details of organelle shapes and structures. Biochemistry- and genetic-based approaches then enabled a detailed molecular framework to be superimposed onto observed subcellular structures. This work was complemented with immuno-based localization studies at the light and electron microscopy level, pinpointing the steady-state distributions of involved proteins. At that time, in the 1980s and 1990s, available techniques provided only limited insight into the dynamic nature of subcellular events. The introduction of video/time-lapse microscopy to monitor micro-injected fluorescently tagged molecules in living cells bridged this gap and greatly enriched our understanding of dynamic events at a cellular level.

At the turn of the millenium, imaging techniques such as Förster resonance energy transfer (FRET) and fluorescence lifetime imaging (FLIM) were used to visualize phosphorylation-based signaling in a landmark publication (Verveer et al. 2000). This was followed by the development of specialized probes known as biosensors cleverly designed to reveal signal transduction events at the subcellular level (see, e.g., Machacek et al. 2009). The introduction and use of the green fluorescence protein (GFP) in the

Correlative Imaging: Focusing on the Future, First Edition. Edited by Paul Verkade and Lucy Collinson.
© 2020 John Wiley & Sons Ltd. Published 2020 by John Wiley & Sons Ltd.

mid-1990s enabled the visualization of protein dynamics at the subcellular and molecular level using fluorescence recovery after photobleaching (FRAP) and fluorescence correlation spectroscopy (FCS). This made it possible to determine the subcellular position of fluorescent molecules including their steady-state mobility; i.e., the fraction of molecules that are mobile and their rates of diffusion. Combined, the early 2000s saw a dramatic increase in the use of fluorescence-based microscopy, mainly using GFP and its variants, and other fluorescent proteins.

The growth in the use of fluorescence microscopy techniques continues even today with new probes and advanced instruments being developed to probe ever deeper into cellular events. What has remained a challenge throughout is the ability to link an observed cellular event to its correct ultrastructural context. Ideally, this should be done using an observed fluorescence probe to precipitate chemical compounds, such that these can be viewed later at the ultrastructural level in the same section. This requires the same cell and subcellular structure to be identified (correlated) at the resolution of light (e.g. fluorescence) and that of the electron beam (e.g. EM). As such, techniques have been developed enabling the same cell and subcellular structure to be viewed at multiple magnifications (Figure 2.1). Commonly, the emitted fluorescence is used to generate free radicals directly or in combination with the use of peroxidase to precipitate a chemical

(a)

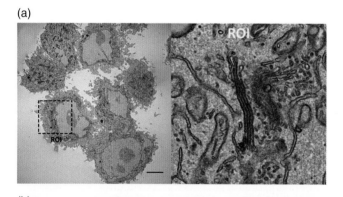

(b)

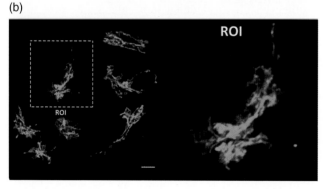

Figure 2.1 Golgi apparatus in Hela Cells; (a) Transmission electron micrograph; and, (b) Laser scanning confocal microscopy image of Golgi-resident glycosylation enzyme, N-acetylgalactosaminyltransferase-2 fused to green fluorescent protein. Enlarged regions of interest (ROI) are shown in the 2 right panels. scale bar 5 μm

compound to create an electron dense footprint that can then be viewed by EM. This combination of light and electron microscopy is termed correlative light and electron microscopy (CLEM), and it affords the dual advantages of fluorescent imaging (spatial and temporal) with the high-resolution of electron microscopy. The overlay of fluorescence and electron microscopy images enables us to determine the subcellular architecture including any morphological and/or ultrastructural changes induced by the molecule of interest. Such dual examination provides invaluable complementary and often unique information. Bridging the gap between light and electron microscopy, however, is challenging from both a technical and logistical perspective, the latter stemming from the fact that light and electron microscopy have been developed separately, rarely crossing paths.

2.1.2 Correlative Microscopy: Two Cultures Collide

The development of light microscopy at the turn of the seventeenth century profoundly expanded our understanding of both life and material sciences, and subsequently enhanced our ability to diagnose disease and identify disease-relevant agents providing early insight into pathogens such as malaria and tuberculosis. The revolution in qualitative biological science, stimulated by the development of the first microscopes roughly 400 years ago, was followed more recently by a revolution in quantitative science stimulated by the development of digital imaging. Scientists could not take advantage of their full potential until digital imaging made it possible to leverage computational tools for image analysis, including powerful computers, fast and sensitive high-resolution digital cameras, and improved image processing software, revealing a new understanding of the complexity of biological systems (Tsien 2003). Researchers can now use automated microscopy systems to mine thousands of cell images in a single experiment. Currently, bioimaging is having its second spark in revolutionizing life science, providing means for direct visualization and quantification of intracellular events in cells and tissues, probing the structure and function of bio-molecules.

The instruments used for microscopy, and workflows needed for sample preparation, are in constant development, all aimed at providing new biological information. This includes improved resolution, sensitivity, and 3D imaging capabilities with enhanced temporal and spatial resolution for live-cell imaging. We now appreciate that cellular structures are both complex and dynamic, that they encompass many structural variations, and that biomolecules may be involved in more than one process at the same time (e.g. clathrin coat components). A comprehensive understanding of cellular processes therefore requires multiple biomolecules to be imaged at the same time, at a resolution comparable to their size, and importantly, with a reference to their particular subcellular environment (i.e. their ultrastructural context). As yet, no single instrument enables this, and there is consequently a growing desire for correlative methodologies that can bridge datasets obtained by different imaging techniques. CLEM provides this, while enabling researchers to gain additional and often important morphological information. The ultrastructural reference provided through CLEM should, in principle, enable multiple sets of imaging modules to be carried out. There is no doubt that these combined imaging methods have become the standard methodology in biological imaging as there is an increasing need to close the spatial and temporal resolution gap between routine microscopy observations and high-resolution multidimensional reconstructions over length scales ranging from micrometers to nanometers.

There are four main reasons for applying CLEM: (i) to complement dynamic information from live cell imaging with high-resolution ultrastructural information; (ii) to provide information on the exact location or state of a specific fluorescently labelled molecule within the ultrastructural context; (iii) to enable by light microscopy the screening of a complex and heterogeneous sample to identify rare events that are difficult or impossible to distinguish by electron microscopy; and (iv) to gain a reference after multi-modal imaging. A merging of the two cultures of light and electron microscopy is required to enable CLEM to be used to its full potential.

2.2 Microscopy Multiculturalism

2.2.1 When Fluorescence Light Microscopy Resolution Is Not Enough

For decades, electron and light microscopy have been developed separately. Although microscopy experts might have the same goal of imaging samples at resolutions beyond the human eye, how each one accomplishes that can be quite different: each workflow has specialized equipment, sample preparation and fixation requirements, software packages, and applications. As a result, most researchers are schooled and trained in one technique, the one that's best suited to solve their exact research question. This is perhaps best exemplified by the cultural divide that exists between light and electron microscopy dating back more than half a century. This separation is further cemented by the housing of equipment used for light microscopy, usually well away from any electron microscope. More often than not, they are located in separate buildings, departments, or even faculties across the campus.

Some institutes don't even have electron microscopes, despite having advanced imaging platforms. This not only makes transfer of samples from one system to another challenging or impossible but also ensures that the two remain isolated from each other, including instrumentation, knowhow, and defining user needs for new developments. This separation is not limited to academia; multiple cultures exist within microscope manufacturers as well. For most companies, the light and electron microscopy branches of the business have two distinct organizations within the same company. Each have evolved products, software platforms, and expertise separately. Thankfully, this is about to change as modern research requires implementation of approaches and instruments to enable viewing at multiple levels of resolution. In this respect, development of CLEM is steadily advancing and a large number of recent strategies that combine light and electron microscopy are already being applied to unravel multiple aspects of cell biology that have not been amenable by other means. Also, the establishment of robust protocols and routines to perform correlative experiments has facilitated the application of CLEM by less experienced users.

Nevertheless, CLEM still remains a demanding technique and is perhaps best conducted through centralized microscopy platforms that maintain up-to-date knowhow, ensure access to modern instruments, and promote development of new techniques, software, and hardware. As the switch between the resolution of the light wave and the electron is rather large, we predict that CLEM, particularly in the context of viewing ultrathin tissue sections, will also incorporate reflective contrast and super-resolution microscopy as intermediate steps.

2.2.2 The Fluorescence Microscopy (FM), Needle/Haystack Localization

FM is a powerful tool for observing specific molecular components in living cells, despite its relatively low spatial resolution. FM remains a universal technique in the life sciences, in which fluorescence emission from a specimen can be readily viewed. In this technique, the sample absorbs incident photons and emits light (signal) at a distinct, wavelength longer than that of the incident photons (Figure 2.2a). Hence, it is particularly powerful because of its high specify, sensitivity, and ability to contrast different parts of labeled samples. In particular, the development of Nobel Prize–winning concept of genetically expressed fluorescent labels, such as green fluorescent protein and its various variants, has produced a step change for biomedical imaging (Chalfie et al. 1994). This has been applied to label specific genes, molecular complexes, and also proteins in order to study their localizations, functionality and interactions within live cells. Immunofluorescent labeling approaches, in which the specificity of antibodies to their antigen is employed, also had a profound effect on our understanding the cellular structure and function (Figure 2.2). However, in fluorescence microscopy only the labeled molecules of interest can be detected and visualized, while contextual information on the rest of cellular architecture is missing.

2.2.3 Electron Microscopy, Visualizing the Ultrastructure

EM, contrary to diffraction limited FM, has among the highest resolution of imaging techniques yet cannot provide temporal information in living cells. Since its first

(a)   (b)

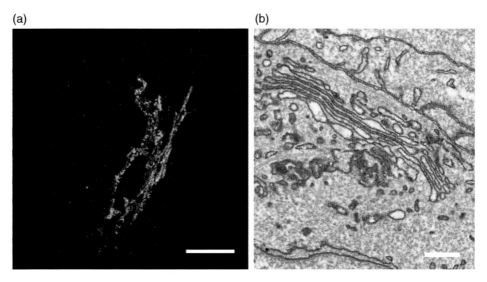

Figure 2.2 High resolution imaging of the Golgi apparatus in Hela Cells; (a) LSM 880 Airyscan image of fixed and processed for indirect immunofluorescence: Golgi-resident enzyme, Galactose-1-Phosphate Uridyltransferase (green) and gp27 (TMEM7, red), scale 5 μm; and, (b) scanning electron micrograph image taken by array tomography, scale 200 nm.

implementation by Ruska and Knoll in 1932, it has served as the main tool for ultra-structural high-resolution imaging in biological sciences as well as in physical sciences. The EM provides a whole ultra-structural map of the sample, delivering comprehensive information on the substructure of intracellular compartments, but often with no molecule specificity (Figure 2.2b). Despite the ultrastructural information and incomparable resolution that it provides in near-native state, EM procedures can only be applied to fixed biological specimen, unlike in non-invasive FM. Fixation methods also profoundly affect both ultrastructural preservation and antigenicity of epitopes within the sample. Staining and labeling opportunities for simultaneous visualization of several different proteins are much narrower compared to specific labeling possibilities in FM. These limitations make EM a rather complementary method to FM in the life sciences.

2.2.4   Finding Coordinates

The principle of CLEM is straightforward: Observe and document the specimen at the light microscopy level, and then retrace the particular cell or area imaged at the ultrastructural level by EM. In reality, the process is demanding with respect to skill. Correlative light and electron microscopy is manual, tedious, and time-consuming at best. It's an attempt to image the same region of a cell or tissue using two different instruments with very high precision, to an accuracy of a micron or less. And researchers have been doing this essentially blindfolded, with few labels or markers that are compatible with both techniques to guide the way. With the introduction of optical subdiffraction imaging methods (Hell 1994; Betzig et al. 2006), however, scientists began using light microscopy to localize single molecules in 3D with a resolution approaching that of the electron microscope. However, live-cell imaging does not reveal membrane morphology and context. Conversely, electron microscopy provides a higher resolution and a more complete context of a cell's structure and morphology while being restricted to fixed cells. It cannot image living material. In short, fluorescence microscopy provides images of light signals on a black background on the microscale, while electron microscopy provides grayscale images of cellular components on the nanoscale.

In a typical correlative microscopy workflow, researchers first use fluorescence microscopy to observe an event on the cellular level. The sample is then fixed and transferred to an electron microscope to get the specific information about the structure surrounding that event. This allows researchers to scan the cell or tissue for interesting regions with fluorescence microscopy and then "zoom in" for a more detailed perspective with electron microscopy. Because sample preparation for electron microscopy includes chemical fixation and heavy-metal stains that quench fluorescence, the order of the workflow is almost never reversed.

However, the major problem is retracing the observed event at the EM level: that is, moving a sample from one instrument to another and then finding the exact same region at a greatly increased magnification, with different imaging additives and with different outputs! In the early days of correlative microscopy, this required researchers to hand-draw coordinate systems based on some internal markers in a sample under a microscope. This was achieved through the use of structural features, introduction of electron-dense nanoparticles, simple scratches in the glass or plastic surface, or by using imaging dishes coverslips that contain an "embossed finder pattern" that is visible under both light and electron illumination. Then the researcher would attempt to find those markers under the next microscope at a lower resolution, and then slowly zoom in from there.

Finally, to correlate super-resolution images with electron microscopy, high precision becomes even more important. Since we work with images in the nanometer range, the coordinate registration and tracking must be more accurate than when using conventional fluorescent microscopy images. Hence, with super-resolution single-molecule localization imaging, we need to make sure the molecule that we visualize by EM is the one that is producing the fluorescent signal and not a neighboring molecule.

2.3 Bridging the Gap between Light and Electron Microscopy

2.3.1 Finding the Same Cell Structure in Light and Electron Microscopes

The wavelength of accelerated electrons is much shorter than visible light, so the diffraction barrier can be overcome and smaller features visualized. But resolution is not the only difference between fluorescence and electron microscopy. There are also differences in the type of contrast typically measured. Whereas FM can generally image specifically labeled macromolecules, EM provides primarily contextual information. In life science, EM is used to analyze membrane structures such as the endoplasmic reticulum, the Golgi apparatus, vesicular structures and so on. CLEM illustrates the sequential combination of light and electron microscopy and represents a versatile tool to correlate functional with structural information from a singular biological event. Correlative microscopy was originally developed to bridge the gap between light and electron microscopy, to provide ultrastructural context in fixed cells cell (Wetzel et al. 1973). This was followed by CLEM in living cells to study sperm motility setting the stage for correlative microscopy. As fluorescent proteins became mainstream, CLEM saw its application in the field of membrane traffic revealing tubular connections between membrane-bound cisternae of the Golgi apparatus as possible transport intermediates (Polishchuk et al. 2000). The experimental scheme is to prepare a fusion construct, comprising a fluorescent protein in one of the resident protein markers of a transport organelle (e.g. the Golgi complex or the endoplasmic reticulum), and then determine its location, structure, and cellular context (e.g. vicinity to mitochondria or endosomes) of these endomembranes as well as of the membranes originating from it, such as transport carriers, by fluorescence microscopy. Fluorescence microscopy is thus the essential step that defines the process under study and clarifies which stages of the carrier's life must be analyzed at higher resolution by EM. And the second step is the determination the ultrastructure of the moving fluorescent carrier at predefined stages of its lifecycle (e.g. carrier formation or fusion) by EM. Conceptually, it can be seen as a switch of wavelength, from that of photons to that of electrons, with a corresponding huge gain in resolution and key information pertaining to ultrastructural context, a feat that not even super-resolution light microscopy can provide.

2.3.2 Making the Fluorescence Labels Visible in the Electron Microscope

One of the most successful methods that allow the same stained specimen to be used for both fluorescence and electron microscopy is fluorescence photooxidation, developed initially by Maranto (1982), later extended to fluorescent proteins by us and other groups (Grabenbauer et al. 2005; Gaietta et al. 2006), and to APEX oxidation, a genetically encodable peroxidase not dependent on light activation (Martell et al. 2012). The photo-oxidation method relies on the fact that when certain fluorescent dyes are excited in the presence of diaminobenzidine (DAB), the reactive oxygen produced by the triplet-excited fluorescent compound causes the DAB to form a deposit very close to the reaction site. This deposit can be stained with considerable specificity with osmium tetroxide (Figure 2.2). Because the reaction is limited to only the region near the excited fluorescent dye, the precise area can be located in the thin epoxy-embedded section by electron microscopy. As a result, this correlative imaging approach allows the researcher to gain additional novel morphological information and this provides a degree of confidence about the structures of interest, as information obtained with one method can be directly compared to that seen with the other methods.

2.3.3 Visualizing Membrane Trafficking Using CLEM

There is growing interest in visualizing and characterizing the vesicular and tubular components of membrane trafficking. Due to their size, usually in the order of 60 nm, vesicles such as COPI or COPII appear as points in the fluorescence microscope, lacking any ultrastructural and morphological detail. Our group has developed a correlative photooxidation method that builds on earlier correlative studies that used concentrated light sources of particular wavelengths to generate an electron-dense reaction product by photooxidation, a highly localized precipitate specific to the structure of interest (Grabenbauer et al. 2005). Our method, termed GRAB for GFP Recognition After Bleaching, allows for the direct ultrastructural visualization of GFP upon illumination and uses oxygen radicals generated during the GFP bleaching process to photooxidize 3,3′-diaminobenzidine (DAB) into an electron-dense precipitate that can be visualized by routine transmission electron microscopy and electron tomography (Figure 2.3). This ability, to bring together correlative labelling with time-resolved microscopy represents a powerful tool for future studies to dissect membrane traffic to, through and from the Golgi.

As GFP and related variants are so commonly used on cell biology, we focused on developing conditions in which the free radicals generated by GFP can be harnessed to precipitate DAB. We found, that the DAB precipitate appears to be locally distributed, indicating a close spatial relationship between the chromophore (GFP) and the final precipitate. The main goal was to establish conditions that produce a DAB precipitate of sufficient quality to permit high-resolution investigations. Visualization in three dimensions is also necessary to extract spatial and temporal information of fluorescent fusion proteins in highly convoluted membrane structures such as those involved on the Golgi architecture. As established Golgi markers, we used the human Golgi-resident glycosylation enzyme, N-acetylgalactosaminyltransferase-2 (GalNAc-T2) fused to green or cyan enhanced fluorescent protein (Figure 2.3a and b). The GRAB technique offers a method to quantitatively, correlate fluorescence to protein distribution at the ultrastructural level. In conclusion, the possibility of performing correlative and quantitative illumination-base electron microscopy with fluorescent proteins and their variants is and will be of great value for the community working with live cells as it offers a direct route from fluorescence to the ultrastructural level.

(a) (b)

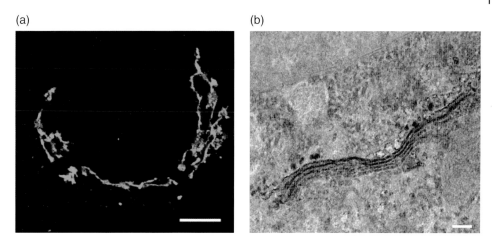

Figure 2.3 Photooxidation of GFP polymerizes DAB to an electron-dense precipitate. (a) Super-resolution structured illumination microscopy shows the Golgi-resident glycosylation enzyme, GalNAcT2^{GFP} stably expressed in HeLa cells, scale 5 µm. (b) After photooxidation and epoxy embedding, the DAB staining can be identified by transmission electron microscopy resembling the GFP localization, scale bar 200 nm.

2.4 Future CLEM Applications and Modifications

2.4.1 Correlative Reflection Contrast Microscopy and Electron Microscopy in Tissue Sections

The bridging of multiple levels of resolution extends back to the early 1960s where researchers compared sections through histochemistry and immunohistochemistry through light microscopy with that of parallel sections at the EM level. As peroxidase is commonly used to define antigens through indirect immuno-labeling, the same enzyme and reaction can be used to precipitate electron-dense substrates visible at the EM level (Liposits et al. 1986). This even enables double-labeling experiments to be carried out (Joosten 1990). A great improvement when performing correlative (immuno) histochemistry is achieved through the application of reflection contrast microscopy (RCM) introduced by Curtis in 1964 and Ploem in 1975. As the authors have elegantly demonstrated, RCM is truly ideal for viewing thin tissue sections as this greatly improves contrast (Cornelese-ten 1988; Cornelese-ten and Velde 1990). This "intermediate" step in resolution also serves as a useful bridge between light and electron microscopy, greatly aiding in the identification of correct cellular structures. As with cell biology-based CLEM, viewing of tissue sections through RCM/CLEM requires ease of handling and the merging of two separate disciplines, pathology and microscopy. For this to happen in the clinic, a new generation of pathologists will have to be trained and willing to incorporate advanced imaging techniques for the benefit of the patient. There is no real reason why this should not be possible, as other advanced imaging techniques are now mainstream (e.g., MRI, PET). In addition, RCM/CLEM has the potential to incorporate mass spectrometry-based techniques such as MALDI imaging (Chaurand et al. 2006) and TOF-SIMS (Nygren et al. 2004) to view molecules at the cellular and subcellular level, respectively. Here, serial sections are used to superimpose RCM/CLEM-based information with that of MALDI- and TOF-SIMS-based imaging. Both enable viewing of molecules such as triglycerides and cholesterol and can also be used to detect small proteins as well as protein-derived tryptic peptides. Importantly, both techniques can also be used to delineate borders between healthy and diseased cells with unprecedented precision thereby providing invaluable information to the pathologist.

2.4.2 Dynamic and Functional Probes for CLEM

An important strength of light microscopy is the ability to directly observe dynamic events in live cells, including cell migration and the translocation of subcellular organelles. It is even possible to directly image the dynamics of signal transduction by using FRET. This is a process involving the radiation-less transfer of energy from a "donor" fluorophore to an "acceptor" fluorophore under certain conditions, including the maximum physical separation of 5 nm between the two fluorophores. FRET enables the quantitative analysis of molecular dynamics in biophysics and in molecular biology, such as the monitoring of protein-protein interactions, protein-DNA interactions, and protein conformational changes. FRET-based biosensors have been utilized to monitor cellular dynamics not only in heterogeneous cellular populations, but also at the single-cell level in real time. Such biosensors are engineered to change conformation in response to activation of a signaling pathway. This conformational change alters the distance between two fluorophores within the biosensor, which can be read out as a change in FRET. Applications of FRET-based biosensors range from basic biological to biomedical disciplines. Despite the diverse applications of FRET, these probes still suffer from the relatively poor spatial resolution and lack of ultrastructural detail needed to fully understand the details of signal transduction in many systems.

There are two general approaches to FRET biosensor design. In the first approach, a full-length signaling protein is incorporated along with an effector domain from a second protein having affinity for the signaling protein in its active state. This approach was first used by Matsuda to develop probes for Ras and Rap GTPases (Mochizuki et al. 2001) and later expanded to include Rho-family GTPases Rac, Cdc42, and Rho (Itoh et al. 2002; Yoshizaki et al. 2003). Many dozens of such probes now exist (reviewed in (Kiyokawa et al. 2011)). The second approach is primarily useful for kinases and involves combining a peptide phosphorylation sequence with a protein domain having differential affinity for the phosphorylated sequence. This approach has been used to target a wide variety of protein kinases including PKC (Violin et al. 2003), Src (Wang et al. 2005), ERK (Harvey et al. 2008), and many others.

FRET biosensors have other useful properties, which extend beyond their ability to dynamically respond to changes in cell signaling. For instance, such probes are genetically engineered and therefore eliminate the need for production of antibodies which specifically recognize active and inactive forms of signaling molecules. This enables detection of any GTPase, even ones for which good antibodies to the active state do not exist. FRET probes are also generally amenable to fixation using a variety of methods, in contrast to many antibodies. This contributes to a third useful property of FRET probes: the same probe can be easily used across different model systems, from cells in culture, to 3D models, to developmental and mouse models (Johnsson et al. 2014; Nobis et al. 2017; Timpson et al. 2011). This greatly simplifies the interpretation of results and is an especially appealing feature for drug discovery, where a typical pipeline involves developing probes and assays to characterize drug activity first in cell culture and later in animal models (Conway et al. 2014). Assays which work in cell culture must often be abandoned or redeveloped for use in animal models, for example due to the use of different fixation methods. When new assays for the same drug activity are developed for animal models, a disconnect between two assays is introduced. Results from one assay must be calibrated against the new one, which introduces uncertainty. FRET biosensors can alleviate this problem, as the same biosensor can be used in vitro and in vivo.

How might this idea be adapted for use in electron microscopy? FRET biosensors depend on substantial conformational changes within the probe to alter the distance between donor and acceptor fluorophores. The same conformational changes could be used to conceal or expose epitope tags for electron microscopy. For example, a hypothetical biosensor might also incorporate a polyhistidine tag at the C- or N-terminus in addition to an internal FLAG-tag (DYKDDDDK, Figure 2.4). In the inactive state, both tags are visible and could be detected using immune-gold labeling with 5 and 10 nm particles (Figure 2.4, a and c). Activation and folding of the biosensor would obscure availability of the FLAG-tag (Figure 2.4,b and c). If needed, this effect could be modulated through the use of additional protein domains; including bulky groups to probe steric hindrance of antibody access to the tag, or electrostatic domains to otherwise interfere with antibody binding. The geometry of FRET biosensors is highly modular (Hodgson et al. 2008), and could be adapted to achieve optimal concealment of the internal epitope in the active state. Such probes would complement the light

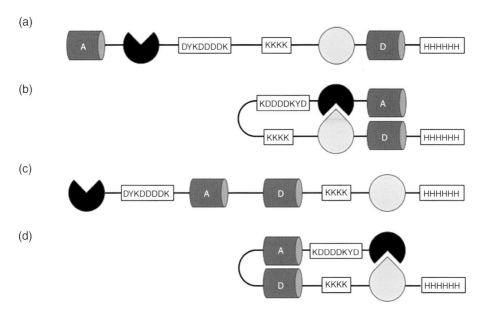

Figure 2.4 Proposed scheme for adaptation of FRET biosensors for use in electron microscopy. In the inactive, open conformation both FLAG and His tags are accessible (a and c). Following activation (b and d) the biosensor folds up due to a change in affinity between functional regions (black and grey circles). Folding of the biosensor brings FRET acceptor and donor fluorophores together (blue barrels) and masks the FLAG tag. Double antibody labelling of the probe thus indicates inactivity whereas single labelling of the probe indicates signal activity.

microscopical use of biosensors to study the dynamics of signal-transduction by enabling high-resolution determination of the exact location and/or compartment of
active signals using EM.

References

Betzig E, Patterson GH, Sougrat R, Lindwasser OW, Olenych S, Bonifacino JS, et al.
Imaging intracellular fluorescent proteins at nanometer resolution. Science; 2006
Sep 15; 313(5793):1642–5.

Chalfie M, Tu Y, Euskirchen G, Ward WW, Prasher DC. Green fluorescent protein as a
marker for gene expression. Science; 1994 263: 802–805.

Chaurand P, Norris JL, Cornett DS, Mobley JA, Caprioli RM. New developments in
profiling and imaging of proteins from tissue sections by MALDI mass spectrometry.
J Proteome Res.; 2006 Nov;5(11):2889–900.

Cornelese-ten Velde I, Prins FA. New sensitive light microscopical detection of colloidal
gold on ultrathin sections by reflection contrast microscopy. Combination of reflection
contrast and electron microscopy in post-embedding immunogold histochemistry.
Histochemistry; 1990;94(1):61–71.

Cornelese-ten Velde I, Bonnet J, Tanke HJ, Ploem JS. Reflection contrast microscopy.
Visualization of (peroxidase-generated) diaminobenzidine polymer products and its
underlying optical phenomena. Histochemistry; 1988 89(2):141–50.

Conway JR, Carragher NO, and Timpson P. Developments in preclinical cancer imaging:
innovating the discovery of therapeutics. Nat Rev Cancer; 2014 14:314–28.

Curtis AS. The mechanism of adhesion of cells to glass. A study by interference reflection
microscopy.J Cell Biol.; 1964 Feb;20:199–215.

Gaietta GM, Giepmans BN, Deerinck TJ, Smith WB, Ngan L, Llopis J, et al. Golgi twins in
late mitosis revealed by genetically encoded tags for live cell imaging and correlated
electron microscopy. Proc Natl Acad Sci USA; 2006 103:17777–82.

Harvey CD, Ehrhardt AG, Cellurale C, Zhong H, Yasuda R, Davis RJ, and Svoboda K. A
genetically encoded fluorescent sensor of ERK activity. Proc Natl Acad Sci USA; 2008
105:19264–19269.

Hell SW, Wichmann J. Opt Lett; 1994 19:780–82.

Hodgson L, Pertz O, and Hahn KM. Design and optimization of genetically encoded
fluorescent biosensors: GTPase biosensors. Methods Cell Biol.; 2008 85:63–81.

Itoh, RE, Kurokawa, K, Ohba, Y, Yoshizaki H, Mochizuki N, and Matsuda M. Activation of
rac and cdc42 video imaged by fluorescent resonance energy transfer-based single-
molecule probes in the membrane of living cells. Mol Cell Biol.; 2002 22:6582–91.

Johnsson, AE, Dai Y., M. Nobis, M.J. Baker, E.J. McGhee, S. Walker, et al. The Rac-FRET
mouse reveals tight spatiotemporal control of Rac activity in primary cells and tissues.
Cell Rep.; 2014 6:1153–64.

Joosten EA. An ultrastructural double-labelling method: immunohistochemical
localization of cell adhesion molecule L1 on HRP-labelled developing corticospinal tract
axons in the rat. Histochemistry; 1990 94(6):645–51.

Kiyokawa, E, Aoki K, Nakamura T, and Matsuda M. Spatiotemporal regulation of small
GTPases as revealed by probes based on the principle of Forster Resonance Energy
Transfer (FRET): Implications for signaling and pharmacology. Annu Rev Pharmacol
Toxicol.; 2011 51:337–58.

Liposits Z, Sherman D, Phelix C, Paull WKA. A combined light and electron microscopic immunocytochemical method for the simultaneous localization of multiple tissue antigens. Tyrosine hydroxylase immunoreactive innervation of corticotropin releasing factor synthesizing neurons in the paraventricular nucleus of the rat. Histochemistry; 1986 85(2):95–106.

Machacek, M., L. Hodgson, C. Welch, H. Elliott, O. Pertz, P. Nalbant, et al. Coordination of Rho GTPase activities during cell protrusion. Nature; 2009 461:99–103.

Maranto AR. Neuronal mapping: a photooxidation reaction makes Lucifer yellow useful for electron microscopy. Science; 1982 Sep 3;217(4563):953–55.

Martell JD, Deerinck TJ, Sancak Y, Poulos TL, Mootha VK, Sosinsky GE, et al. Engineered ascorbate peroxidase as a genetically encoded reporter for electron microscopy. Nat Biotechnol.; 2012 Nov;30(11):1143–48.

Mochizuki N, Yamashita S, Kurokawa K, Ohba Y, Nagai T, Miyawaki A, et al. Spatio-temporal images of growth-factor-induced activation of Ras and Rap1. Nature.; 2001 411:1065–68.

Nobis, M, Herrmann D, Warren SC, Kadir S, Leung W, Killen M. et al. A RhoA-FRET Biosensor Mouse for Intravital Imaging in Normal Tissue Homeostasis and Disease Contexts. Cell Rep.; 2017 21:274–88.

Nygren H[1], Malmberg P, Kriegeskotte C, Arlinghaus HF. Bioimaging TOF-SIMS: localization of cholesterol in rat kidney sections. FEBS Lett.; 2004 May 21;566(1–3):291–23.

Ploem JS. Reflection-contrast microscopy as a tool for investigation of the attachment of living cells to a glass surface. In: von Furth R, editor. Mononuclear Phagocytes in Immunity, Infection and Pathology. Melbourne, London: Blackwell Scientific Publications; 1975. pp. 405–21. Classic work on applying IRM imaging to biological samples using epi-illumination.

Polishchuk RS, Polishchuk EV, Marra P, Alberti S, Buccione R, Luini A, et al. Correlative light–electron microscopy reveals the tubular–saccular ultrastructure of carriers operating between Golgi apparatus and plasma membrane. J Cell Biol; 2000 148(1): 45–58.

Timpson, P, McGhee EJ, and Anderson, KI. Imaging molecular dynamics in vivo – from cell biology to animal models. J Cell Sci.; 2011 124:2877–90.

Tsien, RY. Imagining imaging's future. Nat. Rev. Mol. Cell Biol.; 2003 S, S16–S21.

Verveer PJ, Wouters FS, Reynolds AR, Bastiaens PI. Quantitative imaging of lateral ErbB1 receptor signal propagation in the plasma membrane. Science; 2000 Nov 24;290(5496): 1567–70.

Violin JD, Zhang J, Tsien RY, and Newton AC. A genetically encoded fluorescent reporter reveals oscillatory phosphorylation by protein kinase C. J Cell Biol.; 2003. 161:899–909.

Wang, Y, Botvinick EL, Zhao Y, Berns MW, Usami S, Tsien RY, et al. Visualizing the mechanical activation of Src. Nature; 2005 434:1040–45.

Wetzel B, Erickson BW Jr., Levis WR. The need for positive identification of leukocytes examined by SEM. Proc. of Sixth Annual Scanning Electron Microscopy Symposium, IITRI, Chicago, IL; 1973, pp. 535–42.

Yoshizaki H, Ohba Y, Kurokawa K, Itoh R.E, Nakamura T, Mochizuki N, et al. Activity of Rho-family GTPases during cell division as visualized with FRET-based probes. J Cell Biol.; 2003 162:223–32.

3

The Importance of Sample Processing for Correlative Imaging (or, Rubbish In, Rubbish Out)

Christopher J. Peddie[1] and Nicole L. Schieber[2]

[1] Electron Microscopy Science Technology Platform, The Francis Crick Institute, London, United Kingdom
[2] Cell Biology and Biophysics Unit, European Molecular Biology Laboratory, Heidelberg, Germany

3.1 Introduction

Allow us, if you would, to set the scene, deep within the bowels of a once glossy space age structure. Take a walk down the stairs, pass swiftly around the dimly lit corner, under that relentlessly flickering fluorescent light, dodge past the ancillary room where pumps wheeze and chillers whine, and scoot down the corridor paved with cracked and faded linoleum to take your seat, hunched over in the darkness with hands placed left and right, before a faintly illuminated green screen. Allow your eyes to become accustomed to the darkness, and as you begin to navigate this luminous projection of an almost entirely alien landscape, the passage of electrons through your laboriously prepared ultrathin serial section reveals a multitude of strange and intriguing structures. As time gently passes by and viewing continues, both the lengthy sample preparation and the smashing of kneecaps on the bottom of the microscope column are forgiven, and you start to become completely absorbed in the process of ultrastructural analysis, recording your precious findings image by image in dark seclusion. With that, the backdrop for our discourse is set, so let's now get on with the serious business of science.

Given the number of times the following statement has been made in recent years it is becoming somewhat of a cliché, but it's worth repeating for the time being at least, that electron microscopy (EM) has undergone a quiet revolution over the past few decades, both in terms of resolution and scale (Caplan et al. 2011; Collinson and Verkade, 2015; Koning et al. 2018; Kremer et al. 2015; McDonald, 2009; Peddie and Collinson, 2014; Titze and Genoud, 2016). It is no longer the stereotyped world of old that we alluded to above, comprised solely of the many levels of grayscale needed for interpretation of cell ultrastructure. Well, all right, you've got us there. It is certainly still grayscale. But, it's now possible to almost routinely take that detailed grayscale structural information and place some multidimensional fluorescence data on the top.

Enter from stage left, correlative light and electron microscopy. Correlative. Light. Electron. Microscopy. CLEM. In its purest form, CLEM can be used to examine the

Correlative Imaging: Focusing on the Future, First Edition. Edited by Paul Verkade and Lucy Collinson.
© 2020 John Wiley & Sons Ltd. Published 2020 by John Wiley & Sons Ltd.

location and identity of an individual fluorescent structure, and extends to the localization of dynamic, unique events within heterogeneous cells and/or tissues. Three main forces drive CLEM-specific developments: sample processing, technological advances, and ultimately, of course, the push to answer biological questions. The diversity of CLEM methods seen today is a product of the push and pull of these forces. Indeed, research laboratories are increasingly recognising the potential of CLEM methods in answering the questions that are inaccessible, using light microscopy alone or with other methods; and if you particularly enjoy a challenge, you may even wish to study a process that is rare both temporally *and* spatially.

Sample processing is crucial for enabling the transition between the correlative imaging methodologies that allow us to study these very different functional and structural landscapes. Ask someone who has worked in the field for any length of time what the most important part of electron microscopy is, and you'll probably hear a chorus of 'sample preparation', followed by a persistent echo of 'rubbish in, rubbish out'. Despite the stated prerequisite for high quality, though, it is surprising that sample processing still remains something of a mystical art, that thing between the fresh sample and the final image that is largely ignored, remaining in the shadows while other more glamorous innovations bask in the limelight. Nobody does it the same way, and consequently, there are few truly standardized workflows. Nevertheless, there are some reference works, the go-to publications that describe many of the available options (e.g., Griffiths, 1993; Hayat, 1970; Hayat, 1993; McDonald, 2009; McDonald, 2014). The principles of sample preparation follow a pattern of specimen immobilization and preservation, contrasting (where necessary), and stabilization. Generalizing somewhat, two different avenues are possible: room temperature, and cryogenic temperature. Each has key advantages and disadvantages, and a lot of crosstalk is possible between the two, which can make it difficult to rigidly assign methods to either category.

Diving into the details, we can break these processes down further into several equally important steps that must be accomplished before preparation for a specific imaging technique. First, we need to immediately stop the cells in their tracks, and immobilize each and every internal process by fixation using either chemicals that crosslink macromolecules, or rapid freezing in vitreous ice. For plastic embedding in conventional, or part-cryogenic hybrid workflows, some degree of electron contrasting can be added using cocktails of heavy metals, before stabilizing the samples for electron imaging under vacuum by embedding in resin. We've all seen enough sci-fi to have at least some appreciation of what might happen to us if exposed to the vacuum of space, so it's possible to imagine what will happen to a cell exposed to the vacuum needed for electron imaging. But, these resins do not mix with water, so an exchange must be negotiated, first swapping all the water in the sample with a solvent and then swapping the solvent with a resin which is polymerized either chemically at high temperature or with ultraviolet light at low temperature.

Fully cryogenic workflows keep the sample embedded in a layer of vitreous ice throughout, though keeping the samples well frozen throughout a given workflow can sometimes be a challenge. The ice itself provides rigidity, electron contrast when coupled with advanced detection methods, and stability under vacuum (Carroni and Saibil, 2016). Sample geometry is very restricted, at around 10 microns in depth for plunge frozen material (Dubochet et al. 1988), to around 200 microns for most high pressure frozen samples (McDonald, 2009). Larger samples can be cryoprotected and plunge frozen (Tokuyasu, 1973), which also permits immunolabeling (good!), but membrane contrast is compromised (bad!). Chemical fixation with resin embedding can move us into the range of 1 mm³ specimens (see also Bleck et al. (2009) for comparison of some of these methods). The main point here is that there are many different approaches out there, each with positive and negative aspects; so, depending on the complexity of the question, sample type, and how ambitious you want to be (subject also to the skillset of your local friendly EM community), the best approach may turn out to be just one, or multiple interconnected methods.

Weaving established electron microscopy methods into correlative workflows requires careful consideration in order to balance the demands of each modality without too much compromise, while matching imaging pipelines that at first glance can appear to be completely incompatible. Our main focus here will be to explore some of the possibilities for next generation sample processing in the context of current correlative workflows, but we'll also touch on some related technology developments. For instance, the ability to directly correlate between imaging modalities is helping to fuel a resurgence of interest in ultrastructural microscopy techniques. Coupled with improvements in electron imaging technologies, it is one of the forces behind the seismic shift from two-dimensional to three-dimensional representation. Many of the biological electron microscopy projects that were traditionally the bread and butter work of the transmission electron microscope (TEM), have now transferred to scanning electron microscopy (SEM) based modalities, which, coupled with three-dimensional (3D) light microscopy, are more amenable to large scale volume-based studies of what are of course naturally 3D structures. For higher resolution work, both at room and cryogenic temperatures, the TEM still sits imperiously at the top of the pyramid, enabling examination of and correlation to structures and macromolecular complexes at subnanometer scales.

The introduction and development of serial blockface scanning electron microscopy (SBF SEM; Denk and Horstmann, 2004) is an example that demonstrates the forces of sample preparation, technological advances, and biological question working together. When it emerged, this 3D imaging technique was immediately recognized for application in large-scale mapping studies and its potential for contribution to what has now become termed connectomics, the comprehensive mapping of neuronal networks, down to the level of individual synaptic contacts (Briggman and Bock, 2012; DeFelipe, 2010; Eberle et al. 2015; Helmstaedter, 2013; Lichtman et al. 2008; Morgan and Lichtman, 2017; Swanson and Lichtman, 2016; Wanner et al. 2015). Though applicability was initially held back by limitations in sample preparation, the potential impact of the technique was widely recognized and this pushed forward rapid developments that enhanced electron contrast and improved the stability of specimens (Deerinck et al. 2010; Hua et al. 2015; Mikula and Denk, 2015; Starborg et al. 2013; Tapia et al. 2012), some of which were just old things rediscovered in the bright light of a new technology, and others in refinement of the technology itself. Thus, SBF SEM has enabled three-dimensional (3D) visualization of a range of samples, a previously laborious and very time-consuming task (Titze and Genoud, 2016; White et al. 1986). Continuing improvements have catapulted the technique into the widespread applications we see today, and having got the sample processing refined, have seen it transfer into correlative applications as a powerful localization tool (Bushong et al. 2015; Lees et al. 2017; Russell et al. 2017). This pattern of development can be seen across all forms of electron microscopy, from the simplest of TEM methods to the most complex of volume electron microscopy methods, and many other new developments in sample processing and technology are now coming to light.

3.2   Searching for Correlative Electron Microscopy Utopia

As we've outlined already then, high-quality sample preparation is, above all else, critical for successful correlative microscopy. There. That's all you really need to know— the rest of the chapter encapsulated in a single line. Job done. Oh no, wait. *Specifically,* what is important for correlative processing, and what might the future hold for electron microscopy sample preparation? Well now, there's a can of worms. Indeed, you may well ask, isn't that like trying to look into a crystal ball for answers (Richard Webb, personal communication)? And, you may be correct. But, staring deeply into the aforementioned crystal ball, what might we consider as one very idealistic vision of the future for correlative sample processing?

An optically transparent conductive resin used to rapidly embed specimens with cubic centimeter dimensions containing contrast resistant dyes with identified, imaged, and marked regions of interest with known coordinates that can be automatically tracked through each imaging modality to produce a fully overlaid and precisely aligned correlated 3D volume at less than 5 nanometer isotropic scale that can then be automatically processed to segment and render both the specific features of interest and the surrounding structure with extraction of relevant statistical parameters. That's quite a mouthful. Plus, it would require some extremely careful work to come up with a suitably catchy acronym for the workflow (because, as we all know, a recognizable, somewhat ironic yet catchy acronym is the first and most important step in the development of any new method or instrument. And the compulsory addition of ome and/or omics in the headline process. Don't forget the omics).

Realistically, will this happen any time soon? No, probably not. It's a very complicated jigsaw, and many of the pieces are either not ready or don't quite fit yet. But at the time of writing, some of the parts needed for this type of workflow are rapidly developing, so it could actually be closer than expected. Let us now explore some of the future possibilities for correlative sample processing in the context of this idealistic target and review the big headlines, the main act in our little opus: how can CLEM be made more accessible, smarter, and more automated? How will sample preparation play an integral role in the correlative workflows of the future?

3.3   Sample Processing for Correlative Imaging: A Primer for the First Steps

Higher, faster, stronger. Or, if you prefer, *citius, altius, fortius.* It's not just the spin behind the Olympic spirit, but is also similar to the trends underlying many of the driving forces behind advances in instrumentation — the endless quest for larger fields of view, greater sensitivity, improved spatial resolution, and faster acquisition times. Sample preparation has by necessity had to keep up as imaging methodologies evolve, and the research questions of interest become more complex; in this context it isn't just about processing the specimen through to a plastic block, but also extends into working with the specimen to suit the requirements of multiple imaging techniques. We're not going to cover every possibility here; there simply isn't the space or the time, so we'll try to be brief and just quickly introduce the basics in the context of correlative processing (besides which, if you aren't already set for an afternoon nap, you might be by the time we get to the end).

It is of course the biological question of interest that determines the first steps in sample preservation for CLEM, and dictate the path that must be followed for downstream processing; these are choices that seem likely to stay with us for the foreseeable future. Briefly examining the signposts at this first fork in the road to CLEM, there are a multitude of complex options and target specific considerations to look at, many of which overlap. How large is the sample? How small is the target area? What resolution is needed to address the question of interest? Does the biological process occur over a long or short time scale? Do we work at room temperature, or should we use cryogenic techniques? Will the structure survive processing intact? Will processing artefacts influence the biological interpretation? How will the region of interest be relocated? For what length of time must the fluorescent signal be retained? Casting your gaze further down the path, the forest of signposts that pops into view can be quite intimidating.

If you're going to work toward the ultimate in sample preservation and resolution, cryo-techniques are preferred in comparison to chemical fixation (McDonald 2009; McDonald 2014; Weston et al. 2009), and they're often considered the gold standard for many applications. Fully cryogenic applications exist in an almost entirely separate world, requiring specific strategies for correlation, with special sample holders for imaging and transfers between microscopes (Briegel et al. 2010; Jun et al. 2011; Koning et al. 2014; Li et al. 2018; Rigort et al. 2012a; Schorb and Briggs, 2014; Schorb et al. 2017; Schwartz et al. 2007; van Driel et al. 2009). Because sample sizes are restricted, the specimen and area of interest must either start out thin enough, for example at the cell periphery, or be made thin enough for electron imaging (Al-Amoudi et al. 2007; Hayles et al. 2010; Mahamid et al. 2015; Rigort et al. 2012b; Villa et al. 2013). Correlative cryo-fluorescence and cryo-electron microscopy on compatible specimens, offers formidable benefits in terms of throughput, accuracy and performance (Briegel et al. 2010; Wolff et al. 2016; Zhang 2013). The proof, as they say, is in the pudding, revealing its true forte in the high-resolution study of specific subcellular structures and molecules (Hampton et al. 2017; Jun et al. 2011; Koning et al. 2014; Sartori et al. 2007; Schellenberger et al. 2014; Schwartz et al. 2007; van Driel et al. 2009). There is no question of preserving fluorescent markers; vitrify your sample properly, and the signal will almost always be maintained. Fluorescence lifetime is also enhanced under cryogenic conditions, which is advantageous for imaging (Le Gros et al. 2009; Liu et al. 2015; Schwartz et al. 2007) and lends itself well to super-resolution imaging (Chang et al. 2014; Liu et al. 2015; Wolff et al. 2016).

Reconstructing whole cells using electron tomography, though, is challenging (Noske et al. 2008). Sacrificing ultrastructural resolution, vitrified adherent cells can be studied intact using cryo-correlative soft X-ray tomography (cryo-CLXM), which is advantageous for studying larger volumes and allows the cellular environment to be mapped in a more complete fashion (Carzaniga et al. 2014; Dent et al. 2014; Duke et al. 2014). Should higher resolution be required, these approaches are also compatible with later examination by cryo-EM, or processing for conventional EM studies. It is also important to highlight, before we embark on discussion of room temperature methods (and ignoring for now the cheerleaders with their chants of *cryo, cryo, CRYO!*), that comparisons between cryo-SXT and volume EM of whole cells have shown that the differences in structure are often not enough to influence the underlying biological interpretation (Muller et al. 2012).

Room temperature CLEM, on the other hand, while perhaps considered less technically demanding, has a more diverse range of applications. As a result, there can be more levels to navigate in the decision tree, and it is to this realm that we now shift our emphasis. A major concern to address in relation to correlative imaging is this: when do you want to carry out your fluorescence imaging? This is one of the points at which CLEM begins to expand, and so long as consideration is taken to image that fluorescence at the correct time during processing, without compromising ultrastructural preservation, there can be many downstream options. In practical terms, this often comes down to a first choice of retaining fluorescence after resin embedding, or not. To retain fluorescence, hybrid methods featuring parts of both cryogenic and room temperature techniques represent powerful options, such as after high-pressure freezing and freeze substitution (Johnson et al. 2015; Kukulski et al. 2011; Nixon et al. 2009; Peddie et al. 2014); more on this later. If fluorescence does not need to be retained, the best possible ultrastructural preservation is usually assured, and conventional processing methods that follow the general scheme we outlined earlier can be applied, with the proviso that some kind of map is definitely going to be needed in order to navigate back to the region of interest; more on this later as well.

3.4 Making It Go Faster (We Want More Speed, More Speed…)

Traditional bench processing protocols for chemical fixation and embedding, and freeze substitutions, can be lengthy exercises requiring oodles of patience and careful timing. So, would you believe us if we told you that it's not really necessary for EM sample preparation to take days or weeks at a time? It may seem incredible, but you can even reduce the total processing time to just a few hours! We're not pulling your leg, it's not a joke, it really can be that fast! Three specific examples are microwave (MW) assisted processing, quick-freeze substitution (QFS), and rapid resin infiltration and polymerization.

Microwave ovens were first shown to stabilize tissue by Mayers (1970). It was later shown that microwaves, in combination with chemical fixatives, greatly decreased fixation times (Login and Dvorak 1985; Zimmerman and Raney 1972), but it took another 12 years for electron microscopy MW tissue processing to gain renewed attention. Early on, it was difficult to obtain reproducible results. A turning point came when Giberson et al. (1997) published a protocol showing MW-assisted processing within 4 hours that was reproducible and comparable to conventional methods. Later, the implementation of a vacuum chamber within the MW drastically reduced sample preparation times (~30% compared to other MW protocols, and more than 90% when compared to conventional bench protocols; Giberson and Demaree 2001).

The first commercial laboratory MW processor designed specifically for histopathology and electron microscopy was the Bio-Rad H2500, introduced in the 1980s and later re-released by Polaron Equipment Ltd. (now Quorum Technologies Ltd.). Many other models have since been released as the technology improves. The underlying principle of operation differs slightly for each, but key to their success and reproducibility are variable power controllers that supply continuous power output within a specific range, and precise temperature control to restrict heating of the sample (Giberson and Demaree 2001; Zechmann and Zellnig 2009).

Research into the mechanisms behind MW-assisted processing, alongside technological developments, have brought this method to routine use in many EM labs (Giberson and Demaree 1995; Giberson and Demaree 2001; Giberson et al. 1997; Kok and Boon 2003; Kok et al. 1994; Leong and Sormunen 1998; Login and Dvorak 1994a; Login and Dvorak 1994b). MW processors can take the specimen all the way from primary fixation to polymerization, or in any other combination with conventional bench protocols, as well as many other applications. The dielectric fields induced by MW energy cause dipolar molecules like water to rapidly oscillate (at a rate of some 2.45 billion cycles per second!) through 180 degrees, which increases inter- and intra-molecular motion, and thermal energy (Leong and Sormunen 1998; Login and Dvorak 1985). This enables the rapid diffusion of fixatives and stains into specimens and enhances chemical reactions (Kok and Boon, 1990; Login and Dvorak 1994b). In conjunction, vacuum exposure is thought to alter tissue dynamics and improve diffusion rates over shorter time scales (Russin and Trivett 2001). Throughout the history of MW processing, there has been much debate over the thermal and nonthermal effects of MW irradiation (Galvez et al. 2004). Tinling et al. (2004) directly correlated MW radiation with reduced decalcification times to support the idea that low-wattage MW radiation facilitates diffusion independently of temperature. This notion is also supported by the common use of the ColdSpot® from Ted Pella Inc., a device that actively dissipates excess heat to ensure an even MW distribution that is nonthermal and thus relies solely on the mixing effects of increased molecular motion (Giberson and Demaree 2001; Wendt et al. 2004).

MW irradiation not only reduces the time taken to prepare samples for EM, but has also been shown to improve ultrastructural preservation in TEM sample preparations on retina (Wendt et al. 2004), zebrafish larvae (Schieber et al. 2010), Pacific yew needle, and mouse liver and kidney (Giberson and Demaree 1995). At the very least, the quality is directly comparable to conventional methods (Giberson et al. 1997; Schieber et al. 2010; Zechmann and Zellnig 2009). Despite this positive evidence in different model organisms and tissues, it is not a universally accepted method, and many labs still prefer to use conventional bench processing protocols.

Why should this be? If you can speed up what is, let's face it, an already pretty tedious task by incorporating some new equipment, wouldn't you grab at it with both hands? Your viewpoint may depend on which school of EM you attended, but modern MW processing systems are capable of producing consistently excellent sample preparations in a fraction of the time, so what's not to like? Perhaps this can be partly explained by highlighting that electron microscopists in general tend to be a really stubborn bunch; the processing is complex, and there are a lot of places where it can go slightly wrong, so if we've got something that works well, it can take a very long time to build up the momentum and enough of a consensus to make even a small change. The finer points very rarely match between laboratories, or even between individuals within the *same* laboratory! Microwave processors of course aren't the be all and end all, and we're not advocating their use in all applications. Successful sample preparation can obviously be achieved in many different ways, and in the end the methods used are more sample dependent than anything else, but for many they offer very clear advantages and can really increase throughput.

In the same year that it was suggested that microwaves could speed up the time spent to process samples for EM, Hayat and Giaquinta (1970) demonstrated that embedding could be completed in as little as 4 hours. Work by Hawes et al. (2007), and later McDonald and Webb (2011), showed that freeze substitution doesn't need to take place over extended time frames. Instead, it can be completed within 3 hours for quick freeze substitution (QFS) or even 90 minutes for the super quick version (SQFS), with extremely nice structural preservation (McDonald and Webb 2011). By comparison, most conventional freeze substitution protocols fall in the range of 2–4 days for an epoxy resin embedded sample, and 7 days or more for an acrylic resin embedded sample. It is also possible to speed up epoxy resin infiltration using centrifugation, and polymerize more quickly by increasing the oven temperature to 100°C, allowing blocks to be cured in as little as 2 hours (rather than the longstanding convention of 48 hours), without any compromise to ultrastructure or sectioning quality (McDonald 2014). The freeze substitution process itself can also be microwave enhanced (Hansen et al. 2010). If you combine QFS or SQFS with MW-enhanced infiltration and increased temperature curing, it becomes possible to make it from live specimen to the electron microscope in a single day!

3.5 Embedding Resins

Resins are a surprisingly complex topic, and there are many factors to consider for each electron imaging modality. At present, a one-size-fits-all approach doesn't work. The most commonly used resin formulations, epoxy and acrylic/methacrylate, each have advantages and disadvantages. These bias toward some applications and away from others, but are less than perfect in each individual scenario, which leaves some room for improvement. The specific characteristics of each resin determine which is the most appropriate for a given application. Examples of factors to think about include sectioning properties, stability under the electron beam, preservation of antigenicity, preservation of fluorescence, and viscosity.

For correlative applications in particular, the behaviour of sections and blocks during exposure to the electron beam is an important consideration. If we are to accurately match and overlay data from multiple imaging modalities, there must be as little change as possible in the sample when recording these images. In most cases, the changes that occur during dehydration and embedding (such as uneven infiltration, warping, and most significantly, shrinkage) will outweigh those that occur during imaging (such as distortion of TEM sections during electron beam exposure, or charge induced signal distortion during blockface imaging). Nonetheless, they remain important factors in high-precision correlation.

Epoxy resins are preferred for TEM and array tomography applications. Sections must form stable ribbons, and must be of consistent quality with minimal compression. The sections need to stick to the collection substrate and stay there during any subsequent manipulations (e.g., contrasting or labeling steps). During viewing, especially for tomographic use, the sections must be resistant to radiation damage, mass loss, and shrinkage (Kizilyaprak et al. 2015). They are also preferred for blockface imaging techniques due to their ability to tolerate relatively high cumulative electron doses while retaining good sectioning or milling properties.

The stability of epoxy resins is a result of extensive cross-linking, but this high degree of covalent interaction between resin and biological material presents a problem for immunochemistry applications (Griffiths 1993). For immunolabeling and CLEM therefore, acrylic or methacrylate resins are often preferred. Formulations such as Lowicryl HM20 preserve antigenicity and fluorescence (Nixon et al. 2009; Peddie et al. 2014; Schwarz and Humbel 2007), and if imaging of specimens within the resin is to be considered prior to electron microscopy, optical transparency, and low autofluorescence are prerequisites. These resins also tolerate the addition of low concentrations of water, are low viscosity at low temperatures, allow low temperature polymerization with ultraviolet light, and still retain good sectioning properties and electron beam resistance (Acetarin et al. 1986; Carlemalm et al. 1982; Griffiths 1993).

There have been very few controlled studies of resins for electron microscopy. For instance, Kizilyaprak et al. (2015) compared eight different resin formulations based on the study of Luft (1961), and carried out detailed measurements of performance for TEM and focused ion beam SEM (FIB SEM), concluding that a harder epoxy resin formulation was advantageous in both applications, with a hybrid formulation being the best performing acrylic variant. For such an important component in the imaging pipeline, it remains surprising that there have not been more specific studies in relation to current imaging technologies, especially given the popularity of 3D correlative studies, and this warrants further investigation.

What else can we look out for in the future of resins and embedding? Biological materials are not naturally conductive, and resins are insulating, but during electron imaging, it is necessary to disperse the electrons that are bombarding the sample. In the context of volume EM in particular, 'conductive' resins are spoken of in revered, hushed tones. With the removal of electron beam induced charging artefacts, it could become possible to substantially increase the electron dose that the sample sees during imaging. In turn, this would allow much faster acquisition, and at higher resolution with superior signal-to-noise ratios. But, it's an as-yet-unproven theory. For compatibility with CLEM workflows, it would also be a significant advantage if this magical resin could be optically transparent. Thinking about how we might make a conductive resin, one option is to add something conductive to a standard resin recipe, for instance Nguyen et al. (2016) used a specific type of powdered carbon. But this approach tends to make the resin completely opaque, and this is bad news if you need to be able to see the orientation of your sample after embedding. Other methods to combat charging have included encasing or coating the entire specimen in a secondary conductive material (Titze and Denk 2013; Wanner et al. 2016), or just blowing it away using local gas injection (Deerinck et al. 2018; more on this later). Embed your samples in a transparent but truly 'conductive' or charge dispersing resin, and it is possible that some of the current technical challenges focused around CLEM and volume EM methods could be greatly reduced. Keep your fingers crossed, a level eye to the horizon, and an ear to the ground; it's bound to happen sooner or later.

3.6   Keeping the Region of Interest in Sight

The most common approach to correlate between light and electron microscopy makes use of the fluorescent signal from expressed fluorophores such as those from the almost ubiquitous GFP family. The markers are then tracked between imaging modalities based on physical location. But, during conventional sample preparation they are usually lost, a result of exposure to the combination of fixatives, heavy metal stains, dehydrating agents, and embedding resins. Consequently, the loss of specific positional information for the region of interest is one of the biggest problems faced in correlative workflows. As we scale up in three dimensions, this becomes a much larger spatial problem, so enabling direct coordination between future imaging modalities is crucial. The creation and/or retention of those vital relocation coordinates should therefore be a key consideration in next-generation correlative methods. This is a technically demanding problem that remains largely unsolved in the sense that there is currently no straightforward way to universally fix it for such a wide range of samples and imaging modalities, but it is possible to break the problem down according to scale.

In the absence of fluorescent markers to directly link imaging modalities, relocating the area of interest at the simplest level means using the features of the specimen itself to derive a 2D or 3D roadmap from A to B. For cells cultured in two dimensions, this can be as simple as growing the cells on a marked substrate so that the coordinates can be viewed during light microscopy acquisition, and translated directly into the resin block, e.g., on etched coverslips, or finder grids. If this approach is compatible with your question of interest, then everything is sunny in your world, and congratulations to you. After all, there's absolutely no benefit in making things more complicated than necessary. Alternatively, fiducial markers that are visible in both modalities can be added directly to the sample, such as fluorescent spheres or gold particles, and used in much the same way (Kukulski et al. 2011; Schellenberger et al. 2014). Intrinsic landmarks, such as nuclear or mitochondrial tags, can also be added as part of the primary imaging step, allowing more precise reorientation after embedding and imaging by using the patterns of these larger organelles (Markert et al. 2016; Markert et al. 2017). As sample size increases, however, and with it the three-dimensionality of the structure, the scale of the problem grows and relocation becomes increasingly problematic. Structural landmarks within the sample, such as vasculature and inherent differences in tissue density and cell distribution, become extremely useful (Karreman et al. 2017; Karreman et al. 2016). A couple of specific tools now come into play that can be very effective for helping to relocate that elusive target within much larger samples.

Micro-computed tomography (microCT) is turning out to be a fantastic tool in any EM laboratory. Here, the sample is placed between an X-ray source and detector, rotated, and imaged at multiple different angles. The sequence of images is then back-projected to reconstruct the X-ray absorption at each point within the scanned volume (Metscher 2009a). A nondestructive technique, it is capable of imaging very large fields of view across a broad range of samples, and also bridges the gap between live and processed samples since imaging is possible both fully hydrated, and after resin embedding (or indeed at any point in between). Used as a localization tool, morphological features within opaque specimens can be identified and used to guide ROI relocation. As resolution continues to improve, microCT can only become even more beneficial.

MicroCT can also be used for quality control, identifying defects (Handschuh et al. 2013), and correcting for distortions (Bushong et al. 2015). Ambitious projects focused on whole brain cellular connectomics are making use of it to screen samples before investing in serial section transmission EM (ssTEM) or array tomography projects (Kasthuri et al. 2015; Markram 2012; Mikula 2016). In a CLEM workflow, microCT contributes to all three of our earlier development targets: by linking imaging modalities to improve precision and ROI relocation it makes CLEM smarter (Handschuh et al.

2013; Karreman et al. 2016; Sengle et al. 2013), more accessible (especially with bench top scanners; Metscher 2009a), and more automated (e.g., navigation for multiple FIB SEM acquisitions; Merkle et al. 2014). Though a lack of established contrasting agents initially limited use (Bentley et al. 2007; Johnson et al. 2006; Metscher 2009a; Metscher 2009b), the value of osmium as a differential stain for soft tissue contrasting opened up access to EM processed samples; high atomic number EM stains have correspondingly high X-ray attenuation coefficients which are ideal for microCT imaging (Keene et al. 2014; Sengle et al. 2013). With the realization that a morphological map is actually a really useful reference before embarking for high-resolution imaging, the utility of microCT in improving the accuracy and feasibility of previously difficult CLEM approaches becomes more obvious.

Sengle et al. (2013) first implemented microCT in a correlative imaging workflow comparing adipose tissue in P4 mouse limbs, and were able to confirm that the observed contrast differences corresponded to the underlying structure. Handschuh et al. (2013) detailed a complete correlative workflow, imaging matching regions with MicroCT, light microscopy, and TEM. They expanded on the correlations of Sengle et al. (2013) by demonstrating multiscale visualization of a whole organism, and bridged between imaging modalities using complementary data. Together with Burnett et al. (2014), these studies were instrumental in establishing the power of microCT when applied to EM studies. Karreman et al. (2016) pushed the concept further to gain insights into metastatic events by accurately capturing single tumour cells in the vasculature of the cerebral cortex, and in subcutaneous tumours of mice. Here, microCT was used to perform fast targeting of metastatic events within an EM processed sample prior to imaging by FIB SEM or ssTEM, illustrating the versatility of a powerful correlative workflow in a complex scientific question. Bushong et al. (2015) used microCT in conjunction with diaminobenzidine (DAB) labeling to relocate a specific astrocyte in a mouse brain slice. From these examples, it is clear that using microCT to quickly and accurately locate or relocate regions of interest can substantially increase throughput.

Whereas microCT can be used to get spatial information from an unmodified sample, near-infrared branding of tissue has been employed to directly mark the region of interest in CLEM workflows, thereby providing an obvious landmark for downstream targeting. Essentially, you're painting a big circle around the interesting part of the sample, with a flashing neon arrow pointing straight at it. Hey! Hey! Over here! Over here! Look at me! Look at me! Branding is achieved by using a 2-photon microscope to make marks with micrometer precision within the sample, creating a 3D map around the region of interest. By operating at the near-infrared wavelengths to which tissues are more transparent, deeper penetration is assured, and less damage occurs on the way there because excitation is restricted to the focal plane. Depending on the limitations of the microscope system, and the natural features of the region of interest, these markings can be either simple lines and boxes, or more complex arbitrary shapes (Bishop et al. 2011; Lees et al. 2017; Maco et al. 2013). Asymmetric marks, in particular, are useful because they add another layer of information in terms of position and orientation. Branding can be employed either as a standalone technique (Blazquez-Llorca et al. 2015; Maco et al. 2013), or in combination with MicroCT (Weinhard et al. 2018). In these examples, intact brains or vibratome slices were branded around the region of interest so that the samples could be accurately trimmed and subsequently imaged using FIB SEM with striking accuracy. In this way, individual neurons and processes could be relocated within a mass of otherwise identical neurons. That there is your proverbial needle in a haystack!

3.7   Correlation and Relocation with Dual Modality Probes

Skipping happily forward from relocation using only natively present or introduced physical characteristics, the most common method to correlate between light and electron microscopy relies on fluorescent probes. Ha! You've got us there again; we did already mention that these markers are usually lost during the sample processing steps. But, we're now adding another layer by expanding our focus to include dual modality probes. These are valuable markers for correlative workflows since they make it possible to directly identify the specific region of interest in both imaging modalities, and should remove the uncertainty (and the educated guesswork) in matching the position of fluorophore to structure.

A first option is to work with probes that are independently visible in each imaging modality and therefore allow direct correlation; that is, probes that are both fluorescent, and inherently electron dense. Fluoronanogold (FNG) comprises a ~1.4 nm gold cluster and fluorophore conjugated to an antibody, either as Fab' fragment, IgG, or streptavidin (Cheutin et al. 2007; Powell et al. 1997; Takizawa and Robinson 2000; Takizawa et al. 1998; Takizawa et al. 2015). After fluorescence microscopy, the size of the gold cluster is commonly enhanced further so that it is visible against the cell using either silver or gold metallography for easier EM visibility, though this makes discrimination of multiple labels very difficult because nonuniform enhancement often yields particles of multiple sizes. As the core particle size is much smaller than other secondary immunogold conjugates, however, the better accessibility of FNG to antigens makes for a more sensitive detection system. Quantum dots (QDs) are nanocrystals of differing shape and size that enable multicolor multiplex pre-embedding labeling (Alivisatos et al. 2005; Giepmans et al. 2005). QDs again are smaller than most other immunogold conjugates, so they also offer better penetration and labeling efficiency, but their larger size means that they are slightly less sensitive than FNG. A common limitation for both of these systems is the need to first introduce the probes into the cells/tissues as both are antibody mediated, and this compromises structure, accessibility, and penetration. Put very simply, it's hard to get a big thing (like an antibody conjugate) through a small hole (in the cellular membranes), or indeed a square thing through a round hole, so membrane permeabilization is usually employed to get the antibodies in there. This is normally accomplished using mild detergents, which strips out lipids, altering membrane morphology and causing loss of cytosolic components.

An alternative approach is to work with photoconvertible expressed fluorescent probes. First introduced by Shu et al. (2011), miniSOG (mini singlet oxygen generator) generates singlet oxygen with illumination, and locally generates an electron dense reaction product by reacting with diaminobenzidine (DAB). APEX, a more efficient singlet oxygen generator, was originally designed as a cytosolic alternative to peroxidase labeling (Martell et al. 2012), although this came at the expense of enzyme activity, and in many cases overexpression was required to gain sufficient reactivity with DAB (Lam et al. 2015). With the introduction of APEX2, sensitivity was substantially improved, enabling the detection of endogenous proteins (Lam et al. 2015), and by tagging APEX to a GFP-binding peptide, existing clone libraries can also be exploited, removing the need to remanufacture the marker for each and every application (Ariotti et al. 2015). FLIPPER (fluorescent indicator and peroxidase for precipitation with EM resolution) is another example of a combinatorial CLEM probe (Kuipers et al. 2015). The key difference here was the use of optimized fluorophores (mTurquoise2, EGFP, mOrange2, and mCherry) and a highly active peroxidase. Using these probes, it was possible to distinguish between cell types in a mixed population and examine protein localization (Kuipers et al. 2015). The later development of much smaller FLIPPER-bodies widens applicability and improves the sensitivity of this technology (de Beer et al. 2018). Ongoing work to engineer other FLIPPER-like modules, or APEX variants with even higher sensitivities, underscores the utility of these techniques. These methods though are biased for identification of protein molecules, so what if you are interested in labeling of nonprotein biomolecules? Well, luckily for you, recent developments using 'Click-EM' open up some options here (Ngo et al. 2016; van Elsland et al. 2016). These methods make use of metabolic labeling substrates that are incorporated into cells using the endogenous cellular machinery, and to which singlet oxygen-generating fluorescent dyes can be attached using bioorthogonal click chemistry. Photoconversion then follows, again generating an electron dense reaction product for EM visualization.

Each option has strengths and weaknesses, and to date none has really emerged as a clear favorite. As stated by de Boer et al. (2015), working with fluorescence ensures that no ultrastructural detail is obscured, whereas using electron dense markers (e.g., diaminobenzidine, immunogold) to locally highlight the structure of interest can obscure the finer details. Diffusion of the reaction product from the specific site of photoconversion also limits precision. In the case of expressed genetic probes, no cellular permeabilization is needed, but expression levels must be carefully controlled so as to avoid changing the cellular processes of interest. Photoconversion methods are restricted solely to the specific area of illumination, and some encoded probes have very limited fluorescence emission. For correlative immunoprobes, smaller is better. Probe size versus target size are crucial considerations for the accuracy of correlation at the nanometer scale. For the highest accuracy correlation, coupled with superresolution FM compatibility, probes must be located as close as possible to the genuine site of the molecule of interest. Reduced probe size will also widen the range of possible targets, aid penetration, and remove the requirement for detergent-mediated access.

Monomeric engineered fluorophores (e.g., the mFruit family, are increasingly being used in FM and superresolution imaging methods). As such, they should be incorporated into more dual modality probe constructions. The future evolution of existing probes and derivation of new probes is, by necessity, going to involve a substantial amount of multidisciplinary collaboration. This is a very good thing; it's the best way to quickly and efficiently bring forward new developments, and should be a key focus for future work. And how do we go about making the move from matching two modalities, to correlating between multiple imaging modalities? This is likely to become one of the more prominent areas of development for next generation methods.

3.8 Integration of Imaging Modalities, and In-Resin Fluorescence

How do we get around the need for additional positional information? Do we really need to use indirect tracking methods? What about correlating directly by keeping the fluorescent signal after resin embedding, and adding electron contrast into the mix? Could this be the best of both worlds? Protocols that enable preservation of fluorescence in resin after embedding have been described for a number of model systems, including viruses and yeast (Kukulski et al. 2012a; Kukulski et al. 2011), *C. elegans* (Sims and Hardin 2007; Watanabe et al. 2011), zebrafish (Luby-Phelps et al. 2003; Nixon et al. 2009), *Drosophila* embryos (Fabrowski et al. 2013), plants (Bell et al. 2013), and cultured cells (Bushong et al. 2015; Kukulski et al. 2012b; Peddie et al. 2014). The power of these techniques is that the fluorescent signal, which relates directly back to that previously recorded event, is fully retained and can be used for direct relocation and high precision correlation within exactly the same physical section. It's simply not necessary to add any secondary tracking information because the original fluorescent marker is still there, shining like a beacon. Each of these methods shares some commonality, which clearly demonstrates the key processes, including high pressure freezing, freeze substitution with very low concentrations of stain, and low temperature embedding into acrylic resins with polymerization catalysed by ultraviolet light.

After collection on an EM grid for TEM imaging, or a suitable substrate for SEM imaging, sections from in resin fluorescence (IRF) blocks can be taken forward for fully hydrated fluorescence imaging, before being additionally contrasted for high-resolution electron microscopy. For many fluorophores, enough of the signal is retained to image the fluorescence in semi- and/or ultra-thin sections. The sections can then be imaged for electron microscopy with excellent ultrastructural preservation, allowing the identification of specific sites of endocytosis (Avinoam et al. 2015; Fabrowski et al. 2013; Kukulski et al. 2012a; Kukulski et al. 2011), nuclear pore insertion (Hampoelz et al. 2016), ASC speck formation (Kuri et al. 2017), and intraluminal vesicle dynamics (Adell et al. 2017). This on-section correlative approach uses the best of both modalities by allowing optimized sequential imaging on individual dedicated systems. However, the translation and modification steps that are needed to transfer between imaging modalities reduces the overall accuracy of correlation, and tracking information can still be lost.

The integration of light and electron microscope technologies into a single instrument now becomes particularly relevant, since maintaining emission from embedded fluorescent markers is what makes it possible to work with truly integrated microscope systems (Agronskaia et al. 2008; Faas et al. 2013; Iijima et al. 2014; Kanemaru et al. 2009; Liv et al. 2013; Zonnevylle et al. 2013). While the integration of microscopes was not in itself a new idea (e.g., Wouters et al. (1982; 1985) demonstrated integration into a SEM), it took time for the sample preparation methods needed to fully exploit these technologies to become sufficiently well developed; indeed, it is possible to argue that the technology actually came about before the processing methods were ready for it. A major benefit is very high correlation accuracy, since the specimen is not moved between imaging modalities (Faas et al. 2013; Haring et al. 2017; Zonnevylle et al. 2013). Despite the potential of integrated systems, however, applications have so far remained very limited.

Translating in-resin fluorescence methods in particular for compatibility across a wider variety of sample types has proven to be very difficult; as such, this should be a focus for future work. It is also important to recognize that these techniques come with some very specific compromises. Crucially for resin embedded samples, the freeze substitution conditions necessary to both preserve fluorescence and introduce electron contrast, while optimally preserving and infiltrating the sample as a whole, are incompatible (Peddie et al. 2014). Bias your processing for more electron contrast and fluorescence preservation takes a big hit, but slide back toward perfect fluorescence preservation and electron contrast will be very poor. Together, these two factors mean that a delicate balance between fluorescence and contrast must be reached. It also highlights where work is currently underway to derive the more effective and efficient protocols of the future. For instance, re-engineering cell lines to make use of new contrast resistant probes such as mEos4 could allow much higher concentrations of osmium and uranyl acetate to be used during freeze substitution while maintaining a good level of fluorescence preservation (Paez-Segala et al. 2015). The search for alternative contrasting agents will also have a role to play in future processing methods. Ultimately, a combination of these approaches may yield the most powerful results.

The disparity in resolution between light and electron imaging modalities is another very significant problem when it comes to making direct comparisons. Diffraction limited light microscopy, and nanometer scale electron microscopy simply doesn't add up very well. If we are to address the ease and accuracy of correlation between these extremes, then we need to work with sample preparation methods that enable the collection of higher resolution light microscopy data without compromising structural integrity, and in doing so, make it easier to assign localization to specific subcellular structures. For accessible samples, such as cultured cells and other thin tissues, collecting super-resolution data prior to electron microscopy is not such a big problem. For larger specimens, though, higher-resolution imaging becomes more difficult. Add in to this the very real possibility of artefacts creeping in during processing, including distortion and shrinkage, and collecting the data after stabilization and embedding in resin becomes more appealing. In-resin fluorescence methods can be used to overcome these problems. Physical sectioning removes the limitations in axial resolution, while localization microscopy approaches address lateral resolution limits (Johnson et al. 2015; Kim et al. 2015; Peddie et al. 2017).

3.9 Streamlining the Correlative Approaches of the Future: SmartCLEM

Moving beyond the merging of light and electron imaging modalities down to a single micrograph, in-resin fluorescence techniques open the door to several possibilities for hardware and imaging pipeline development and as a result, next-generation CLEM workflows really could become smarter, and much more efficient. Once you've got your hands on a block containing both fluorescent marker(s) *and* electron contrast, viewing the fluorescent signal directly within the block makes it possible to trim coarsely to the region of interest. Using an epifluorescence microscope to perform this task with greater precision is the next logical step. Two different approaches have been described to date (Brama et al. 2016; Lemercier et al. 2017).

Brama et al. (2016) designed the ultraLM™ as a locator tool to detect and track fluorescent regions of interest within an in-resin fluorescence block as part of a conventional trimming and/or sectioning workflow. In this way, it becomes possible to trim directly to the fluorescent area of interest prior to moving to volume EM techniques such as SBF SEM or FIB SEM, or by smart collection, to pick up serial sections containing only the target area for analysis by ssTEM or array tomography. Lemercier et al. (2017), on the other hand, took a different approach, designing their microtome-integrated-microscope (MIM) for high resolution and high sensitivity imaging of in-resin fluorescence blocks through to sections and grids, translating the specimen into the optical imaging plane for each step. The MIM, coupled with their direct-CLEM workflow, enables extremely selective selection of sections (try reciting that repeatedly in the bar later) and as a result, precisely targeted high-resolution TEM imaging of rare events. In both cases, the downstream imaging steps can be streamlined by focusing only on the areas of specific interest.

Combine these tools with automated sectioning accessories such as the ATUMtome or AutoCUTS add-ons (Li et al. 2017; Schalek et al. 2011), and the fluorescent signal in the sections on the tape could be used to monitor progress through the region of interest. Or, how about generating a 3D fluorescence reconstruction, on the fly, to help guide electron imaging of the matching area? And, if it isn't possible to image the array of sections during collection, this can still be done afterward using a standalone optical imaging system prior to electron imaging. It would, of course, be remiss of us to suggest that in-resin fluorescence techniques are the only way to achieve this aim. Targeted acquisition across an array of sections is equally possible using multiplex labeling strategies (Collman et al. 2015; Micheva and Smith 2007; Micheva et al. 2010; Micheva et al. 2016; Oberti et al. 2011; Simhal et al. 2018), where sections are collected on a suitable substrate, optically imaged, and then contrasted for EM. With the additional development of some software tools to enable automatic detection of cells and sections (Delpiano et al. 2018; Hayworth et al. 2014), it becomes possible to consider automatically screening entire arrays, mapping out each region of interest based on the fluorescent signal, and following up with targeted electron imaging.

3.10   How Deep Does the Rabbit Hole Go?

One of the most powerful solutions, and where SmartCLEM would really demonstrate entitlement to the name, will be the full integration of fluorescence and volume electron imaging (Brama et al. 2016). The incorporation of an optical imaging system directly within the electron microscope edges closer to our earlier utopian vision. At present, very tight spatial constraints within the electron microscope mean that compromises in the sensitivity and field of view for optical imaging are inevitable, and these, in turn, limit the maximum achievable resolution. These are difficult problems, but with some clever innovations, not insurmountable. Add in a few incremental changes in order to fully exploit in-resin fluorescence protocols, and the future possibilities for identifying and tracking regions of interest, targeted acquisition, and automation of electron imaging based directly on feedback from fluorescence imaging, become extremely interesting.

In addition to the improvements that will streamline the imaging pipeline, other potential benefits in a SmartCLEM workflow are substantial reductions in instrument time (since the region of interest can be positively identified and actively tracked) and reduced data footprint (since only data from the specific region of interest is recorded and analysed), with correspondingly increased speed and efficiency (though conversely, by focusing only region of interest, the supplementary data gained around said area that could be mined in later analysis is greatly reduced). Crucially, the preselection of the target area during the acquisition process is a prerequisite to more amenable automation of downstream feature selection, segmentation, and 3D rendering.

Are we moving in the distant future to a setting where electron microscopy facilities comprise multiple anonymous rooms, each of which contains a nondescript black box, marked in large friendly red letters with 'Correlative Electron Microscopy Platform,' with a single opening labeled in the same friendly text 'Insert Sample Here,' and adjacent server address for almost simultaneous data collection? Do we want to? That's a philosophical question of much greater scope than we have the capacity to cover here, Certainly it would lower the entry barriers and blow accessibility to correlative workflows wide open, but let's not forget the other seemingly invisible parts of a successful correlative workflow, including the carefully considered optimization, troubleshooting, quality control, validation, and detailed interpretation. These are all crucial parts of the process, and require skilled intervention, so we're not all out of a job just yet. Phew. Even with the incessant march of technology, it's unlikely that a magic box would be able to cope with this in the short to medium term, but it's an interesting concept to file away for the future.

3.11   Hold That Thought, Though – Is This All Completely Necessary?

With all this talk of enhancing sample processing, it's also important to consider the imaging hardware itself and how changes here can have widespread impact. One specific example is focal charge compensation (FCC), which prevents electron-beam induced charging over the block surface during imaging, one of the most challenging problems faced in SBF SEM. Achieved by addition of a custom designed and very precisely targeted nitrogen gas injection system (Deerinck et al. 2018), the applications of this seemingly simple add-on are immense. Removal of charging-induced artefacts opens the door to higher-resolution imaging, increased imaging speed, and reduced sectioning thicknesses. Perhaps more importantly, it allows data collection from previously difficult samples, and widens accessibility to a broader range of sample preparation techniques that rely on lower levels of staining. For example, Deerinck et al. (2018) demonstrated that it is possible to successfully image historical samples that were prepared for conventional TEM, where the level of staining is very much reduced when compared to standard SBF SEM protocols. With less metal stain, the masking of features is greatly reduced, leading to an arguably truer representation of the underlying structure. Removing the requirement to focus on heavy staining and extreme conductivity could allow processing to veer back toward attaining the best possible ultrastructural preservation from simpler processing procedures, while still benefiting from the greater speed and scale of volume SEM methods.

How does this type of modification fit into future correlative workflows? Simpler processing with less stain, and cleaner structure will of course be extremely useful, and will ultimately improve the precision of correlation. As we discussed earlier, current methods for direct correlation using in-resin fluorescence protocols are also limited by the necessity to balance fluorescence retention against electron contrast with low-level staining. Taking the problem of electron-beam induced charging out of the equation, it may become possible to work more easily with these challenging samples using previously incompatible electron-imaging parameters in the absence of additional staining, and looking to the future has the potential to help, dare we suggest, make multi-modality imaging in three dimensions a routine affair.

3.12   Improving Accessibility to Correlative Workflows

What are some of the factors that make CLEM workflows inaccessible? At a basic level, the cost, perceived complexity, and requirement for specialist equipment have something to answer for; the overall cost of cryogenic methods for example, in comparison to room temperature, is astronomical. Advanced workflows often require advanced instrumentation and complicated sample preparation methods; these are expensive to purchase and run, and for some this may be too big a hurdle. Many of the techniques that we've touched on here also make use of high-end instruments, which limits accessibility to those lucky enough to have all these things in their toy box. But, don't be downhearted; by breaking each workflow down into its component parts, a huge amount can still be achieved using separate basic instruments.

Fortunately, some laboratories are trying to break down these barriers by developing more cost-effective workflows, a very notable example being QFS; wave goodbye to the expensive freeze substitution apparatus, and just use very basic laboratory equipment instead, including polystyrene boxes, dry ice, bubble wrap, and a party lamp (McDonald and Webb 2011). Unfortunately, though, you will still need to keep hold of that high-pressure freezer, for the time being at least. However, the willingness to think around the problem and seek solutions is definitely out there.

What else can be done in the future to widen accessibility? The fact that there is such a wide range of available CLEM techniques can also present a very real barrier. Experienced electron microscopists are needed to make sense of the broad spectrum of options and pinpoint those that will give the best answer to the specifics of the biological question. Books such as CLEM I, II, and III strive to simplify methods by providing clear and precise descriptions of protocols, applications, and the necessary equipment (Müller-Reichert and Verkade 2011; Müller-Reichert and Verkade 2014; Müller-Reichert and Verkade 2017). However, being able to actually apply these methods is still no easy task. On top of the expertise required for decision making, there are often many subtleties in each workflow that are not necessarily covered in detail in the published protocols, and as we touched on earlier, multiple ways to accomplish the same thing. These seemingly minor details can significantly alter outcomes and hamper reproducibility.

Another very relevant point to consider here is that many of the conventional protocols in use today have not changed substantially for decades. Sure, there may have been some very subtle variations introduced over the years, but by and large, the fundamentals haven't changed a bit. What has changed though, is the collection of faces behind the scenes, and this means that an immense wealth of knowledge has been lost, the reasoning behind and understanding of many of the specific steps disappearing in the mists of time. The failure to pass this information down through the generations is a shortcoming of the field. As we alluded to above, sample processing has been ignored to a large extent in the pursuit of other areas of improvement. But, in order to push forward processing developments, we really do need to maintain some knowledge of the *how* and the *why* in order to improve our understanding of the *what* in the next steps to take. What can we do to address this problem in the future, and minimize knowledge loss with the changing of the guard? Those reference works that we mentioned earlier are often necessary in retracing missing steps. One additional possibility, in this age of relentlessly synchronized boxes, drives, clouds, and other myriad online resources, would be to curate a repository of standard operating procedures, detailed methods, and video protocols. The availability of this kind of resource would also help to open accessibility to correlative workflows by making the possibilities more transparent and readily understandable.

The best CLEM methods are often those that laboratories and imaging facilities are able to establish as routine, and these are often disseminated in specialist courses or by visiting the laboratory directly to gain experience of what lies hidden between the lines of the written protocol. To push forward the CLEM workflows of the future (get out those rose tinted spectacles out again!), greater efforts should be made to push forward simplification, reduce costs, enhance reproducibility, and make open collaboration commonplace and more effective.

3.13   Coming to the End

Massively parallel data acquisition is looming just beyond the visible horizon (Doi et al. 2016; Eberle et al. 2015; Kemen et al. 2015; Kruit and Ren 2016; Mohammadi-Gheidari and Kruit 2011; Ren and Kruit 2016). These emerging SEM techniques have the potential to generate vast quantities of data, perhaps several petabytes or more each and every week when operating at the maximum theoretical output speed, which is an extraordinary figure when many institutes don't even own these levels of storage for all of their operations (and keep in mind that this quoted output originates from a single instrument; imagine what a suite of these could accomplish in a single year without even operating close to the theoretical output). How do current CLEM sample preparation methodologies fit into these types of imaging pipeline? The surprising answer is actually very well. These technologies are not at present reliant on the further development of specific methods, but equally, improvement in some areas such as resins and staining will contribute to their success. Building one of these into a fully integrated microscope system would be something to behold.

With these improvements and/or innovations in sample preparation and technologies comes a wider range of possibilities for integration of two or more imaging modalities, and ultimately, usability. It would be misleading to suggest that this makes multi-modality research easier, after all, if it was easy, someone would already be doing it (this being a favourite catchphrase of one of our editors), but it opens up a much wider range of possibilities that can be exploited to answer multiple intriguing biological questions. Making them more accessible, be it through partial automation or through streamlining of both the individual and combined imaging pipelines, will surely continue to push forward the boundaries of discovery. Will the future of sample preparation and correlative electron microscopy really turn out to follow this path in the long run? Probably not, since grandiose predictions rarely turn out exactly the way that was expected, but these are certainly exciting times to be working on such problems. Besides which, it's nice to have something to look forward to, isn't it?

References

Acetarin, J.D., Carlemalm, E., Villiger, W., 1986. Developments of new Lowicryl resins for embedding biological specimens at even lower temperatures. J Microsc 143, 81–88.

Adell, M.A.Y., Migliano, S.M., Upadhyayula, S., Bykov, Y.S., Sprenger, S., Pakdel, M., Vogel, G.F., Jih, G., Skillern, W., Behrouzi, R., Babst, M., Schmidt, O., Hess, M.W., Briggs, J.A.,

Kirchhausen, T., Teis, D., 2017. Recruitment dynamics of ESCRT-III and Vps4 to endosomes and implications for reverse membrane budding. Elife 6.

Agronskaia, A.V., Valentijn, J.A., van Driel, L.F., Schneijdenberg, C.T., Humbel, B.M., van Bergen en Henegouwen, P.M., Verkleij, A.J., Koster, A.J., Gerritsen, H.C., 2008. Integrated fluorescence and transmission electron microscopy. J Struct Biol 164, 183–189.

Al-Amoudi, A., Diez, D.C., Betts, M.J., Frangakis, A.S., 2007. The molecular architecture of cadherins in native epidermal desmosomes. Nature 450, 832–837.

Alivisatos, A.P., Gu, W., Larabell, C., 2005. Quantum dots as cellular probes. Annu Rev Biomed Eng 7, 55–76.

Ariotti, N., Hall, T.E., Rae, J., Ferguson, C., McMahon, K.A., Martel, N., Webb, R.E., Webb, R.I., Teasdale, R.D., Parton, R.G., 2015. Modular Detection of GFP-Labeled Proteins for Rapid Screening by Electron Microscopy in Cells and Organisms. Dev Cell 35, 513–525.

Avinoam, O., Schorb, M., Beese, C.J., Briggs, J.A., Kaksonen, M., 2015. Endocytosis. Endocytic sites mature by continuous bending and remodeling of the clathrin coat. Science 348, 1369–1372.

Bell, K., Mitchell, S., Paultre, D., Posch, M., Oparka, K., 2013. Correlative imaging of fluorescent proteins in resin-embedded plant material. Plant Physiol 161, 1595–1603.

Bentley, M.D., Jorgensen, S.M., Lerman, L.O., Ritman, E.L., Romero, J.C., 2007. Visualization of three-dimensional nephron structure with microcomputed tomography. Anat Rec (Hoboken) 290, 277–283.

Bishop, D., Nikic, I., Brinkoetter, M., Knecht, S., Potz, S., Kerschensteiner, M., Misgeld, T., 2011. Near-infrared branding efficiently correlates light and electron microscopy. Nat Methods 8, 568–570.

Blazquez-Llorca, L., Hummel, E., Zimmerman, H., Zou, C., Burgold, S., Rietdorf, J., Herms, J., 2015. Correlation of two-photon in vivo imaging and FIB/SEM microscopy. J Microsc 259, 129–136.

Bleck, C.K.E., Merz, A., Gutierrez, M.G., Walther, P., Dubochet, J., Zuber, B., Griffiths, G., 2009. Comparison of different methods for thin section EM analysis of Mycobacterium smegmatis. Journal of Microscopy 237, 23–38.

Brama, E., Peddie, C.J., Wilkes, G., Gu, Y., Collinson, L.M., Jones, M.L., 2016. ultraLM and miniLM: Locator tools for smart tracking of fluorescent cells in correlative light and electron microscopy. Wellcome Open Res 1, 26.

Briegel, A., Chen, S., Koster, A.J., Plitzko, J.M., Schwartz, C.L., Jensen, G.J., 2010. Correlated light and electron cryo-microscopy. Methods Enzymol 481, 317–341.

Briggman, K.L., Bock, D.D., 2012. Volume electron microscopy for neuronal circuit reconstruction. Curr Opin Neurobiol 22, 154–161.

Burnett, T.L., McDonald, S.A., Gholinia, A., Geurts, R., Janus, M., Slater, T., Haigh, S.J., Ornek, C., Almuaili, F., Engelberg, D.L., Thompson, G.E., Withers, P.J., 2014. Correlative tomography. Sci Rep 4, 4711.

Bushong, E.A., Johnson, D.D., Jr., Kim, K.Y., Terada, M., Hatori, M., Peltier, S.T., Panda, S., Merkle, A., Ellisman, M.H., 2015. X-ray microscopy as an approach to increasing accuracy and efficiency of serial block-face imaging for correlated light and electron microscopy of biological specimens. Microsc Microanal 21, 231–238.

Caplan, J., Niethammer, M., Taylor, R.M., 2nd, Czymmek, K.J., 2011. The power of correlative microscopy: multi-modal, multi-scale, multi-dimensional. Curr Opin Struct Biol 21, 686–693.

Carlemalm, E., Garavito, R.M., Villiger, W., 1982. Resin development for electron microscopy and an analysis of embedding at low temperature*. Journal of Microscopy 126, 123–143.

Carroni, M., Saibil, H.R., 2016. Cryo electron microscopy to determine the structure of macromolecular complexes. Methods 95, 78–85.

Carzaniga, R., Domart, M.C., Collinson, L.M., Duke, E., 2014. Cryo-soft X-ray tomography: a journey into the world of the native-state cell. Protoplasma 251, 449–458.

Chang, Y.W., Chen, S., Tocheva, E.I., Treuner-Lange, A., Lobach, S., Sogaard-Andersen, L., Jensen, G.J., 2014. Correlated cryogenic photoactivated localization microscopy and cryo-electron tomography. Nat Methods 11, 737–739.

Cheutin, T., Sauvage, C., Tchelidze, P., O'Donohue, M.F., Kaplan, H., Beorchia, A., Ploton, D., 2007. Visualizing macromolecules with fluoronanogold: from photon microscopy to electron tomography. Methods Cell Biol 79, 559–574.

Collinson, L., Verkade, P., 2015. Probing the future of correlative microscopy. Journal of Chemical Biology 8, 127–128.

Collman, F., Buchanan, J., Phend, K.D., Micheva, K.D., Weinberg, R.J., Smith, S.J., 2015. Mapping synapses by conjugate light-electron array tomography. J Neurosci 35, 5792–5807.

de Beer, M.A., Kuipers, J., van Bergen en Henegouwen, P.M.P., Giepmans, B.N.G., 2018. A small protein probe for correlated microscopy of endogenous proteins. Histochem Cell Biol 149, 261–268.

de Boer, P., Hoogenboom, J.P., Giepmans, B.N., 2015. Correlated light and electron microscopy: ultrastructure lights up! Nat Methods 12, 503–513.

Deerinck, T.J., Bushong, E.A., Thor, A., Ellisman, M.H., 2010. NCMIR methods for 3D EM: A new protocol for preparation of biological specimens for serial block-face SEM. Microscopy, 6–8.

Deerinck, T.J., Shone, T.M., Bushong, E.A., Ramachandra, R., Peltier, S.T., Ellisman, M.H., 2018. High-performance serial block-face SEM of nonconductive biological samples enabled by focal gas injection-based charge compensation. J Microsc 270, 142–149.

DeFelipe, J., 2010. From the connectome to the synaptome: an epic love story. Science 330, 1198–1201.

Delpiano, J., Pizarro, L., Peddie, C.J., Jones, M.L., Griffin, L.D., Collinson, L.M., 2018. Automated detection of fluorescent cells in in-resin fluorescence sections for integrated light and electron microscopy. Journal of Microscopy 0.

Denk, W., Horstmann, H., 2004. Serial block-face scanning electron microscopy to reconstruct three-dimensional tissue nanostructure. PLoS Biol 2, e329.

Dent, K.C., Hagen, C., Grunewald, K., 2014. Critical step-by-step approaches toward correlative fluorescence/soft X-ray cryo-microscopy of adherent mammalian cells. Methods Cell Biol 124, 179–216.

Doi, T., Yamazaki, M., Ichimura, T., Ren, Y., Kruit, P., 2016. A high-current scanning electron microscope with multi-beam optics. Microelectronic Engineering 159, 132–138.

Dubochet, J., Adrian, M., Chang, J.J., Homo, J.C., Lepault, J., McDowall, A.W., Schultz, P., 1988. Cryo-electron microscopy of vitrified specimens. Q Rev Biophys 21, 129–228.

Duke, E., Dent, K., Razi, M., Collinson, L.M., 2014. Biological applications of cryo-soft X-ray tomography. J Microsc 255, 65–70.

Eberle, A.L., Mikula, S., Schalek, R., Lichtman, J., Knothe Tate, M.L., Zeidler, D., 2015. High-resolution, high-throughput imaging with a multibeam scanning electron microscope. J Microsc 259, 114–120.

Faas, F.G., Barcena, M., Agronskaia, A.V., Gerritsen, H.C., Moscicka, K.B., Diebolder, C.A., van Driel, L.F., Limpens, R.W., Bos, E., Ravelli, R.B., Koning, R.I., Koster, A.J., 2013. Localization of fluorescently labeled structures in frozen-hydrated samples using integrated light electron microscopy. J Struct Biol 181, 283–290.

Fabrowski, P., Necakov, A.S., Mumbauer, S., Loeser, E., Reversi, A., Streichan, S., Briggs, J.A., De Renzis, S., 2013. Tubular endocytosis drives remodelling of the apical surface during epithelial morphogenesis in Drosophila. Nat Commun 4, 2244.

Galvez, J.J., Giberson, R.T., Cardiff, R., 2004. Microwave Mechanisms — The Energy/Heat Dichoto.

Giberson, R.T., Demaree, R.S., Jr., 1995. Microwave fixation: understanding the variables to achieve rapid reproducible results. Microsc Res Tech 32, 246–254.

Giberson, R.T., Demaree, R.S., Jr., 2001. Microwave Techniques and Protocols Humana Press.

Giberson, R.T., Demaree, R.S., Jr., Nordhausen, R.W., 1997. Four-hour processing of clinical/diagnostic specimens for electron microscopy using microwave technique. J Vet Diagn Invest 9, 61–67.

Giepmans, B.N., Deerinck, T.J., Smarr, B.L., Jones, Y.Z., Ellisman, M.H., 2005. Correlated light and electron microscopic imaging of multiple endogenous proteins using Quantum dots. Nat Methods 2, 743–749.

Griffiths, G., 1993. Fine Structure Immunocytochemistry Springer-Verlag Berlin Heidelberg.

Hampoelz, B., Mackmull, M.T., Machado, P., Ronchi, P., Bui, K.H., Schieber, N., Santarella-Mellwig, R., Necakov, A., Andres-Pons, A., Philippe, J.M., Lecuit, T., Schwab, Y., Beck, M., 2016. Pre-assembled Nuclear Pores Insert into the Nuclear Envelope during Early Development. Cell 166, 664–678.

Hampton, C.M., Strauss, J.D., Ke, Z., Dillard, R.S., Hammonds, J.E., Alonas, E., Desai, T.M., Marin, M., Storms, R.E., Leon, F., Melikyan, G.B., Santangelo, P.J., Spearman, P.W., Wright, E.R., 2017. Correlated fluorescence microscopy and cryo-electron tomography of virus-infected or transfected mammalian cells. Nat Protoc 12, 150–167.

Handschuh, S., Baeumler, N., Schwaha, T., Ruthensteiner, B., 2013. A correlative approach for combining microCT, light and transmission electron microscopy in a single 3D scenario. Front Zool 10, 44.

Hansen, B.T., Dorward, D.W., Nair, V., Fischer, E.R., 2010. Improved Preservation of HeLa Cells by Sequential Chemical Addition During Microwave-assisted Freeze Substitution. Microscopy and Microanalysis 16, 978–979.

Haring, M.T., Liv, N., Zonnevylle, A.C., Narvaez, A.C., Voortman, L.M., Kruit, P., Hoogenboom, J.P., 2017. Automated sub-5 nm image registration in integrated correlative fluorescence and electron microscopy using cathodoluminescence pointers. Sci Rep 7, 43621.

Hawes, P., Netherton, C.L., Mueller, M., Wileman, T., Monaghan, P., 2007. Rapid freeze-substitution preserves membranes in high-pressure frozen tissue culture cells. J Microsc 226, 182–189.

Hayat, M.A., 1970. Principles and Techniques of Electron Microscopy Van Nostrand Reinhold Company.

Hayat, M.A., 1993. Stains and Cytochemical Methods Springer US.

Hayat, M.A., Giaquinta, R., 1970. Rapid fixation and embedding for electron microscopy. Tissue Cell 2, 191–195.

Hayles, M.F., de Winter, D.A., Schneijdenberg, C.T., Meeldijk, J.D., Luecken, U., Persoon, H., de Water, J., de Jong, F., Humbel, B.M., Verkleij, A.J., 2010. The making of frozen-hydrated, vitreous lamellas from cells for cryo-electron microscopy. J Struct Biol 172, 180–190.

Hayworth, K.J., Morgan, J.L., Schalek, R., Berger, D.R., Hildebrand, D.G., Lichtman, J.W., 2014. Imaging ATUM ultrathin section libraries with WaferMapper: a multi-scale approach to EM reconstruction of neural circuits. Front Neural Circuits 8, 68.

Helmstaedter, M., 2013. Cellular resolution connectomics: challenges of dense neural circuit reconstruction. Nat Methods 10, 501–507.

Hua, Y., Laserstein, P., Helmstaedter, M., 2015. Large-volume en-bloc staining for electron microscopy-based connectomics. Nat Commun 6, 7923.

Iijima, H., Fukuda, Y., Arai, Y., Terakawa, S., Yamamoto, N., Nagayama, K., 2014. Hybrid fluorescence and electron cryo-microscopy for simultaneous electron and photon imaging. J Struct Biol 185, 107–115.

Johnson, E., Seiradake, E., Jones, E.Y., Davis, I., Grunewald, K., Kaufmann, R., 2015. Correlative in-resin super-resolution and electron microscopy using standard fluorescent proteins. Sci Rep 5, 9583.

Johnson, J.T., Hansen, M.S., Wu, I., Healy, L.J., Johnson, C.R., Jones, G.M., Capecchi, M.R., Keller, C., 2006. Virtual histology of transgenic mouse embryos for high-throughput phenotyping. PLoS Genet 2, e61.

Jun, S., Ke, D., Debiec, K., Zhao, G., Meng, X., Ambrose, Z., Gibson, G.A., Watkins, S.C., Zhang, P., 2011. Direct visualization of HIV-1 with correlative live-cell microscopy and cryo-electron tomography. Structure 19, 1573–1581.

Kanemaru, T., Hirata, K., Takasu, S., Isobe, S., Mizuki, K., Mataka, S., Nakamura, K., 2009. A fluorescence scanning electron microscope. Ultramicroscopy 109, 344–349.

Karreman, M.A., Ruthensteiner, B., Mercier, L., Schieber, N.L., Solecki, G., Winkler, F., Goetz, J.G., Schwab, Y., 2017. Find your way with X-Ray: Using microCT to correlate in vivo imaging with 3D electron microscopy. Methods Cell Biol 140, 277–301.

Karreman, M.A., Mercier, L., Schieber, N.L., Solecki, G., Allio, G., Winkler, F., Ruthensteiner, B., Goetz, J.G., Schwab, Y., 2016. Fast and precise targeting of single tumor cells in vivo by multimodal correlative microscopy. J Cell Sci 129, 444–456.

Kasthuri, N., Hayworth, K.J., Berger, D.R., Schalek, R.L., Conchello, J.A., Knowles-Barley, S., Lee, D., Vazquez-Reina, A., Kaynig, V., Jones, T.R., Roberts, M., Morgan, J.L., Tapia, J.C., Seung, H.S., Roncal, W.G., Vogelstein, J.T., Burns, R., Sussman, D.L., Priebe, C.E., Pfister, H., Lichtman, J.W., 2015. Saturated Reconstruction of a Volume of Neocortex. Cell 162, 648–661.

Keene, D.R., Tufa, S.F., Wong, M.H., Smith, N.R., Sakai, L.Y., Horton, W.A., 2014. Correlation of the same fields imaged in the TEM, confocal, LM, and microCT by image registration: from specimen preparation to displaying a final composite image. Methods Cell Biol 124, 391–417.

Kemen, T., Malloy, M., Thiel, B., Mikula, S., Denk, W., Dellemann, G., Zeidler, D. 2015. Further advancing the throughput of a multi-beam SEM, pp. 6 SPIE Advanced Lithography, Vol. 9424. SPIE.

Kim, D., Deerinck, T.J., Sigal, Y.M., Babcock, H.P., Ellisman, M.H., Zhuang, X., 2015. Correlative stochastic optical reconstruction microscopy and electron microscopy. PLoS One 10, e0124581.

Kizilyaprak, C., Longo, G., Daraspe, J., Humbel, B.M., 2015. Investigation of resins suitable for the preparation of biological sample for 3-D electron microscopy. J Struct Biol 189, 135–146.

Kok, L.P., Boon, M.E., 1990. Microwaves for microscopy. J Microsc 158, 291–322.

Kok, L.P., Boon, M.E., 2003. Microwaves for the Art of Microscopy Coulomb Press Leyden.

Kok, L.P., Visser, P.E., Boon, M.E., 1994. Programming the microwave oven. J Neurosci Methods 55, 119–124.

Koning, R.I., Koster, A.J., Sharp, T.H., 2018. Advances in cryo-electron tomography for biology and medicine. Ann Anat 217, 82–96.

Koning, R.I., Celler, K., Willemse, J., Bos, E., van Wezel, G.P., Koster, A.J., 2014. Correlative cryo-fluorescence light microscopy and cryo-electron tomography of Streptomyces. Methods Cell Biol 124, 217–239.

Kremer, A., Lippens, S., Bartunkova, S., Asselbergh, B., Blanpain, C., Fendrych, M., Goossens, A., Holt, M., Janssens, S., Krols, M., Larsimont, J.C., Mc Guire, C., Nowack, M.K., Saelens, X., Schertel, A., Schepens, B., Slezak, M., Timmerman, V., Theunis, C.R.V.A.N.B., Visser, Y., Guerin, C.J., 2015. Developing 3D SEM in a broad biological context. J Microsc 259, 80–96.

Kruit, P., Ren, Y., 2016. Multi-Beam Scanning Electron Microscope Design. Microscopy and Microanalysis 22, 574–575.

Kuipers, J., van Ham, T.J., Kalicharan, R.D., Veenstra-Algra, A., Sjollema, K.A., Dijk, F., Schnell, U., Giepmans, B.N., 2015. Flipper, a combinatorial probe for correlated live imaging and electron microscopy, allows identification and quantitative analysis of various cells and organelles. Cell Tissue Res 360, 61–70.

Kukulski, W., Schorb, M., Kaksonen, M., Briggs, J.A., 2012a. Plasma membrane reshaping during endocytosis is revealed by time-resolved electron tomography. Cell 150, 508–520.

Kukulski, W., Schorb, M., Welsch, S., Picco, A., Kaksonen, M., Briggs, J.A., 2011. Correlated fluorescence and 3D electron microscopy with high sensitivity and spatial precision. J Cell Biol 192, 111–119.

Kukulski, W., Schorb, M., Welsch, S., Picco, A., Kaksonen, M., Briggs, J.A., 2012b. Precise, correlated fluorescence microscopy and electron tomography of lowicryl sections using fluorescent fiducial markers. Methods Cell Biol 111, 235–257.

Kuri, P., Schieber, N.L., Thumberger, T., Wittbrodt, J., Schwab, Y., Leptin, M., 2017. Dynamics of in vivo ASC speck formation. J Cell Biol 216, 2891–2909.

Lam, S.S., Martell, J.D., Kamer, K.J., Deerinck, T.J., Ellisman, M.H., Mootha, V.K., Ting, A.Y., 2015. Directed evolution of APEX2 for electron microscopy and proximity labeling. Nat Methods 12, 51–54.

Le Gros, M.A., McDermott, G., Uchida, M., Knoechel, C.G., Larabell, C.A., 2009. High-aperture cryogenic light microscopy. J Microsc 235, 1–8.

Lees, R.M., Peddie, C.J., Collinson, L.M., Ashby, M.C., Verkade, P., 2017. Correlative two-photon and serial block face scanning electron microscopy in neuronal tissue using 3D near-infrared branding maps. Methods Cell Biol 140, 245–276.

Lemercier, N., Middel, V., Hentsch, D., Taubert, S., Takamiya, M., Beil, T., Vonesch, J.L., Baumbach, T., Schultz, P., Antony, C., Strahle, U., 2017. Microtome-integrated microscope system for high sensitivity tracking of in-resin fluorescence in blocks and ultrathin sections for correlative microscopy. Sci Rep 7, 13583.

Leong, A.S., Sormunen, R.T., 1998. Microwave procedures for electron microscopy and resin-embedded sections. Micron 29, 397–409.

Li, S., Ji, G., Shi, Y., Klausen, L.H., Niu, T., Wang, S., Huang, X., Ding, W., Zhang, X., Dong, M., Xu, W., Sun, F., 2018. High-vacuum optical platform for cryo-CLEM (HOPE): A new solution for non-integrated multiscale correlative light and electron microscopy. J Struct Biol 201, 63–75.

Li, X., Ji, G., Chen, X., Ding, W., Sun, L., Xu, W., Han, H., Sun, F., 2017. Large scale three-dimensional reconstruction of an entire Caenorhabditis elegans larva using AutoCUTS-SEM. J Struct Biol 200, 87–96.

Lichtman, J.W., Livet, J., Sanes, J.R., 2008. A technicolour approach to the connectome. Nat Rev Neurosci 9, 417–422.

Liu, B., Xue, Y., Zhao, W., Chen, Y., Fan, C., Gu, L., Zhang, Y., Zhang, X., Sun, L., Huang, X., Ding, W., Sun, F., Ji, W., Xu, T., 2015. Three-dimensional super-resolution protein localization correlated with vitrified cellular context. Sci Rep 5, 13017.

Liv, N., Zonnevylle, A.C., Narvaez, A.C., Effting, A.P., Voorneveld, P.W., Lucas, M.S., Hardwick, J.C., Wepf, R.A., Kruit, P., Hoogenboom, J.P., 2013. Simultaneous correlative scanning electron and high-NA fluorescence microscopy. PLoS One 8, e55707.

Login, G.R., Dvorak, A.M., 1985. Microwave energy fixation for electron microscopy. Am J Pathol 120, 230–243.

Login, G.R., Dvorak, A.M., 1994a. Application of microwave fixation techniques in pathology to neuroscience studies: a review. J Neurosci Methods 55, 173–182.

Login, G.R., Dvorak, A.M., 1994b. Methods of microwave fixation for microscopy. A review of research and clinical applications: 1970-1992. Prog Histochem Cytochem 27, 1–127.

Luby-Phelps, K., Ning, G., Fogerty, J., Besharse, J.C., 2003. Visualization of identified GFP-expressing cells by light and electron microscopy. J Histochem Cytochem 51, 271–274.

Luft, J.H., 1961. Improvements in epoxy resin embedding methods. J Biophys Biochem Cytol 9, 409–414.

Maco, B., Holtmaat, A., Cantoni, M., Kreshuk, A., Straehle, C.N., Hamprecht, F.A., Knott, G.W., 2013. Correlative in vivo 2 photon and focused ion beam scanning electron microscopy of cortical neurons. PLoS One 8, e57405.

Mahamid, J., Schampers, R., Persoon, H., Hyman, A.A., Baumeister, W., Plitzko, J.M., 2015. A focused ion beam milling and lift-out approach for site-specific preparation of frozen-hydrated lamellas from multicellular organisms. J Struct Biol 192, 262–269.

Markert, S.M., Britz, S., Proppert, S., Lang, M., Witvliet, D., Mulcahy, B., Sauer, M., Zhen, M., Bessereau, J.L., Stigloher, C., 2016. Filling the gap: adding super-resolution to array tomography for correlated ultrastructural and molecular identification of electrical synapses at the C. elegans connectome. Neurophotonics 3, 041802.

Markert, S.M., Bauer, V., Muenz, T.S., Jones, N.G., Helmprobst, F., Britz, S., Sauer, M., Rossler, W., Engstler, M., Stigloher, C., 2017. 3D subcellular localization with superresolution array tomography on ultrathin sections of various species. Methods Cell Biol 140, 21–47.

Markram, H., 2012. The human brain project. Sci Am 306, 50–55.

Martell, J.D., Deerinck, T.J., Sancak, Y., Poulos, T.L., Mootha, V.K., Sosinsky, G.E., Ellisman, M.H., Ting, A.Y., 2012. Engineered ascorbate peroxidase as a genetically encoded reporter for electron microscopy. Nat Biotechnol 30, 1143–1148.

Mayers, C.P., 1970. Histological fixation by microwave heating. J Clin Pathol 23, 273–275.

McDonald, K.L., 2009. A review of high-pressure freezing preparation techniques for correlative light and electron microscopy of the same cells and tissues. J Microsc 235, 273–281.

McDonald, K.L., 2014. Out with the old and in with the new: rapid specimen preparation procedures for electron microscopy of sectioned biological material. Protoplasma 251, 429–448.

McDonald, K.L., Webb, R.I., 2011. Freeze substitution in 3 hours or less. J Microsc 243, 227–233.

Merkle, A., Lechner, L., Steinbach, A., Gelb, J., Kienle, M., Phaneuf, M.W., Unrau, D., Singh, S.S., Chawla, N., 2014. Automated correlative tomography using XRM and FIB-SEM to span length scales and modalities in 3D materials. Microsc. Anal. 28, S10–S13.

Metscher, B.D., 2009a. MicroCT for comparative morphology: simple staining methods allow high-contrast 3D imaging of diverse non-mineralized animal tissues. BMC Physiol 9, 11.

Metscher, B.D., 2009b. MicroCT for developmental biology: a versatile tool for high-contrast 3D imaging at histological resolutions. Dev Dyn 238, 632–640.

Micheva, K.D., Smith, S.J., 2007. Array tomography: a new tool for imaging the molecular architecture and ultrastructure of neural circuits. Neuron 55, 25–36.

Micheva, K.D., O'Rourke, N., Busse, B., Smith, S.J., 2010. Array tomography: immunostaining and antibody elution. Cold Spring Harb Protoc 2010, pdb prot5525.

Micheva, K.D., Wolman, D., Mensh, B.D., Pax, E., Buchanan, J., Smith, S.J., Bock, D.D., 2016. A large fraction of neocortical myelin ensheathes axons of local inhibitory neurons. Elife 5.

Mikula, S., 2016. Progress Towards Mammalian Whole-Brain Cellular Connectomics. Front Neuroanat 10, 62.

Mikula, S., Denk, W., 2015. High-resolution whole-brain staining for electron microscopic circuit reconstruction. Nat Methods 12, 541–546.

Mohammadi-Gheidari, A., Kruit, P., 2011. Electron optics of multi-beam scanning electron microscope. Nuclear Instruments and Methods in Physics Research Section A: Accelerators, Spectrometers, Detectors and Associated Equipment 645, 60–67.

Morgan, J.L., Lichtman, J.W., 2017. Digital tissue and what it may reveal about the brain. BMC Biol 15, 101.

Muller, W.G., Heymann, J.B., Nagashima, K., Guttmann, P., Werner, S., Rehbein, S., Schneider, G., McNally, J.G., 2012. Towards an atlas of mammalian cell ultrastructure by cryo soft X-ray tomography. J Struct Biol 177, 179–192.

Müller-Reichert, T., Verkade, P., 2011. Correlative Light and Electron Microscopy. Methods in Cell Biology 111, 1–404.

Müller-Reichert, T., Verkade, P., 2014. Correlative Light and Electron Microscopy II. Methods in Cell Biology 124, 2–442.

Müller-Reichert, T., Verkade, P., 2017. Correlative Light and Electron Microscopy III. Methods in Cell Biology 140, 1–352.

Ngo, J.T., Adams, S.R., Deerinck, T.J., Boassa, D., Rodriguez-Rivera, F., Palida, S.F., Bertozzi, C.R., Ellisman, M.H., Tsien, R.Y., 2016. Click-EM for imaging metabolically tagged nonprotein biomolecules. Nat Chem Biol 12, 459–465.

Nguyen, H.B., Thai, T.Q., Saitoh, S., Wu, B., Saitoh, Y., Shimo, S., Fujitani, H., Otobe, H., Ohno, N., 2016. Conductive resins improve charging and resolution of acquired images in electron microscopic volume imaging. Sci Rep 6, 23721.

Nixon, S.J., Webb, R.I., Floetenmeyer, M., Schieber, N., Lo, H.P., Parton, R.G., 2009. A single method for cryofixation and correlative light, electron microscopy and tomography of zebrafish embryos. Traffic 10, 131–136.

Noske, A.B., Costin, A.J., Morgan, G.P., Marsh, B.J., 2008. Expedited approaches to whole cell electron tomography and organelle mark-up in situ in high-pressure frozen pancreatic islets. J Struct Biol 161, 298–313.

Oberti, D., Kirschmann, M.A., Hahnloser, R.H., 2011. Projection neuron circuits resolved using correlative array tomography. Front Neurosci 5, 50.

Paez-Segala, M.G., Sun, M.G., Shtengel, G., Viswanathan, S., Baird, M.A., Macklin, J.J., Patel, R., Allen, J.R., Howe, E.S., Piszczek, G., Hess, H.F., Davidson, M.W., Wang, Y., Looger, L.L., 2015. Fixation-resistant photoactivatable fluorescent proteins for CLEM. Nat Methods 12, 215–218, 214 p following 218.

Peddie, C.J., Collinson, L.M., 2014. Exploring the third dimension: volume electron microscopy comes of age. Micron 61, 9–19.

Peddie, C.J., Domart, M.C., Snetkov, X., O'Toole, P., Larijani, B., Way, M., Cox, S., Collinson, L.M., 2017. Correlative super-resolution fluorescence and electron microscopy using conventional fluorescent proteins in vacuo. J Struct Biol 199, 120–131.

Peddie, C.J., Blight, K., Wilson, E., Melia, C., Marrison, J., Carzaniga, R., Domart, M.C., O'Toole, P., Larijani, B., Collinson, L.M., 2014. Correlative and integrated light and electron microscopy of in-resin GFP fluorescence, used to localise diacylglycerol in mammalian cells. Ultramicroscopy 143, 3–14.

Powell, R.D., Halsey, C.M., Spector, D.L., Kaurin, S.L., McCann, J., Hainfeld, J.F., 1997. A covalent fluorescent-gold immunoprobe: simultaneous detection of a pre-mRNA splicing factor by light and electron microscopy. J Histochem Cytochem 45, 947–956.

Ren, Y., Kruit, P., 2016. Transmission electron imaging in the Delft multibeam scanning electron microscope 1. Journal of Vacuum Science & Technology B, Nanotechnology and Microelectronics: Materials, Processing, Measurement, and Phenomena 34, 06KF02.

Rigort, A., Villa, E., Bauerlein, F.J., Engel, B.D., Plitzko, J.M., 2012a. Integrative approaches for cellular cryo-electron tomography: correlative imaging and focused ion beam micromachining. Methods Cell Biol 111, 259–281.

Rigort, A., Bauerlein, F.J., Villa, E., Eibauer, M., Laugks, T., Baumeister, W., Plitzko, J.M., 2012b. Focused ion beam micromachining of eukaryotic cells for cryoelectron tomography. Proc Natl Acad Sci U S A 109, 4449–4454.

Russell, M.R., Lerner, T.R., Burden, J.J., Nkwe, D.O., Pelchen-Matthews, A., Domart, M.C., Durgan, J., Weston, A., Jones, M.L., Peddie, C.J., Carzaniga, R., Florey, O., Marsh, M., Gutierrez, M.G., Collinson, L.M., 2017. 3D correlative light and electron microscopy of cultured cells using serial blockface scanning electron microscopy. J Cell Sci 130, 278–291.

Russin, W.A., Trivett, C.L., 2001. Vacuum-Microwave Combination for Processing Plant Tissues for Electron Microscopy, in: R. T. Giberson and R. S. Demaree, Jr., Eds.), Microwave Techniques and Protocols, Humana Press.

Sartori, A., Gatz, R., Beck, F., Rigort, A., Baumeister, W., Plitzko, J.M., 2007. Correlative microscopy: bridging the gap between fluorescence light microscopy and cryo-electron tomography. J Struct Biol 160, 135–145.

Schalek, R., Kasthuri, N., Hayworth, K., Berger, D., Tapia, J., Morgan, J., Turaga, S., Fagerholm, E., Seung, H., Lichtman, J., 2011. Development of High-Throughput, High-Resolution 3D Reconstruction of Large-Volume Biological Tissue Using Automated Tape Collection Ultramicrotomy and Scanning Electron Microscopy. Microscopy and Microanalysis 17, 966–967.

Schellenberger, P., Kaufmann, R., Siebert, C.A., Hagen, C., Wodrich, H., Grunewald, K., 2014. High-precision correlative fluorescence and electron cryo microscopy using two independent alignment markers. Ultramicroscopy 143, 41–51.

Schieber, N.L., Nixon, S.J., Webb, R.I., Oorschot, V.M., Parton, R.G., 2010. Modern approaches for ultrastructural analysis of the zebrafish embryo. Methods Cell Biol 96, 425–442.

Schorb, M., Briggs, J.A., 2014. Correlated cryo-fluorescence and cryo-electron microscopy with high spatial precision and improved sensitivity. Ultramicroscopy 143, 24–32.

Schorb, M., Gaechter, L., Avinoam, O., Sieckmann, F., Clarke, M., Bebeacua, C., Bykov, Y.S., Sonnen, A.F., Lihl, R., Briggs, J.A.G., 2017. New hardware and workflows for semi-automated correlative cryo-fluorescence and cryo-electron microscopy/tomography. J Struct Biol 197, 83–93.

Schwartz, C.L., Sarbash, V.I., Ataullakhanov, F.I., McIntosh, J.R., Nicastro, D., 2007. Cryo-fluorescence microscopy facilitates correlations between light and cryo-electron microscopy and reduces the rate of photobleaching. J Microsc 227, 98–109.

Schwarz, H., Humbel, B.M., 2007. Correlative light and electron microscopy using immunolabeled resin sections. Methods Mol Biol 369, 229–256.

Sengle, G., Tufa, S.F., Sakai, L.Y., Zulliger, M.A., Keene, D.R., 2013. A correlative method for imaging identical regions of samples by micro-CT, light microscopy, and electron microscopy: imaging adipose tissue in a model system. J Histochem Cytochem 61, 263–271.

Shu, X., Lev-Ram, V., Deerinck, T.J., Qi, Y., Ramko, E.B., Davidson, M.W., Jin, Y., Ellisman, M.H., Tsien, R.Y., 2011. A genetically encoded tag for correlated light and electron microscopy of intact cells, tissues, and organisms. PLoS Biol 9, e1001041.

Simhal, A.K., Gong, B., Trimmer, J.S., Weinberg, R.J., Smith, S.J., Sapiro, G., Micheva, K.D., 2018. A Computational Synaptic Antibody Characterization and Screening Framework for Array Tomography. bioRxiv.

Sims, P.A., Hardin, J.D., 2007. Fluorescence-integrated transmission electron microscopy images: integrating fluorescence microscopy with transmission electron microscopy. Methods Mol Biol 369, 291–308.

Starborg, T., Kalson, N.S., Lu, Y., Mironov, A., Cootes, T.F., Holmes, D.F., Kadler, K.E., 2013. Using transmission electron microscopy and 3View to determine collagen fibril size and three-dimensional organization. Nat Protoc 8, 1433–1448.

Swanson, L.W., Lichtman, J.W., 2016. From Cajal to Connectome and Beyond. Annu Rev Neurosci 39, 197–216.

Takizawa, T., Robinson, J.M., 2000. FluoroNanogold is a bifunctional immunoprobe for correlative fluorescence and electron microscopy. J Histochem Cytochem 48, 481–486.

Takizawa, T., Suzuki, K., Robinson, J.M., 1998. Correlative microscopy using FluoroNanogold on ultrathin cryosections. Proof of principle. J Histochem Cytochem 46, 1097–1102.

Takizawa, T., Powell, R.D., Hainfeld, J.F., Robinson, J.M., 2015. FluoroNanogold: an important probe for correlative microscopy. J Chem Biol 8, 129–142.

Tapia, J.C., Kasthuri, N., Hayworth, K.J., Schalek, R., Lichtman, J.W., Smith, S.J., Buchanan, J., 2012. High-contrast en bloc staining of neuronal tissue for field emission scanning electron microscopy. Nat Protoc 7, 193–206.

Tinling, S.P., Giberson, R.T., Kullar, R.S., 2004. Microwave exposure increases bone demineralization rate independent of temperature. J Microsc 215, 230–235.

Titze, B., Denk, W., 2013. Automated in-chamber specimen coating for serial block-face electron microscopy. J Microsc 250, 101–110.

Titze, B., Genoud, C., 2016. Volume scanning electron microscopy for imaging biological ultrastructure. Biol Cell 108, 307–323.

Tokuyasu, K.T., 1973. A technique for ultracryotomy of cell suspensions and tissues. J Cell Biol 57, 551–565.

van Driel, L.F., Valentijn, J.A., Valentijn, K.M., Koning, R.I., Koster, A.J., 2009. Tools for correlative cryo-fluorescence microscopy and cryo-electron tomography applied to whole mitochondria in human endothelial cells. Eur J Cell Biol 88, 669–684.

van Elsland, D.M., Bos, E., de Boer, W., Overkleeft, H.S., Koster, A.J., van Kasteren, S.I., 2016. Detection of bioorthogonal groups by correlative light and electron microscopy allows imaging of degraded bacteria in phagocytes. Chem Sci 7, 752–758.

Villa, E., Schaffer, M., Plitzko, J.M., Baumeister, W., 2013. Opening windows into the cell: focused-ion-beam milling for cryo-electron tomography. Curr Opin Struct Biol 23, 771–777.

Wanner, A.A., Kirschmann, M.A., Genoud, C., 2015. Challenges of microtome-based serial block-face scanning electron microscopy in neuroscience. J Microsc 259, 137–142.

Wanner, A.A., Genoud, C., Masudi, T., Siksou, L., Friedrich, R.W., 2016. Dense EM-based reconstruction of the interglomerular projectome in the zebrafish olfactory bulb. Nat Neurosci 19, 816–825.

Watanabe, S., Punge, A., Hollopeter, G., Willig, K.I., Hobson, R.J., Davis, M.W., Hell, S.W., Jorgensen, E.M., 2011. Protein localization in electron micrographs using fluorescence nanoscopy. Nat Methods 8, 80–84.

Weinhard, L., di Bartolomei, G., Bolasco, G., Machado, P., Schieber, N.L., Neniskyte, U., Exiga, M., Vadisiute, A., Raggioli, A., Schertel, A., Schwab, Y., Gross, C.T., 2018. Microglia remodel synapses by presynaptic trogocytosis and spine head filopodia induction. Nat Commun 9, 1228.

Wendt, K.D., Jensen, C.A., Tindall, R., Katz, M.L., 2004. Comparison of conventional and microwave-assisted processing of mouse retinas for transmission electron microscopy. J Microsc 214, 80–88.

Weston, A.E., Armer, H.E., Collinson, L.M., 2009. Towards native-state imaging in biological context in the electron microscope. J Chem Biol 3, 101–112.

White, J.G., Southgate, E., Thomson, J.N., Brenner, S., 1986. The structure of the nervous system of the nematode Caenorhabditis elegans. Philos Trans R Soc Lond B Biol Sci 314, 1–340.

Wolff, G., Hagen, C., Grunewald, K., Kaufmann, R., 2016. Towards correlative super-resolution fluorescence and electron cryo-microscopy. Biol Cell 108, 245–258.

Wouters, C.H., Koerten, H.K., 1982. Combined light microscope and scanning electron microscope, a new instrument for cell biology. Cell Biol Int Rep 6, 955–959.

Wouters, C.H., Ploem, J.S., 1985. A new instrument combining simultaneous light and scanning electron microscopy. Prog Clin Biol Res 196, 115–133.

Zechmann, B., Zellnig, G., 2009. Microwave-assisted rapid plant sample preparation for transmission electron microscopy. J Microsc 233, 258–268.

Zhang, P., 2013. Correlative cryo-electron tomography and optical microscopy of cells. Curr Opin Struct Biol 23, 763–770.

Zimmerman, G.R., Raney, J.A., 1972. Fast fixation of surgical pathology specimens. Lab. Med. 3, 29–30.

Zonnevylle, A.C., Van Tol, R.F., Liv, N., Narvaez, A.C., Effting, A.P., Kruit, P., Hoogenboom, J.P., 2013. Integration of a high-NA light microscope in a scanning electron microscope. J Microsc 252, 58–70.

4

3D CLEM: Correlating Volume Light and Electron Microscopy

Saskia Lippens[1] and Eija Jokitalo[2]

[1] *BioImaging Core, VIB, Ghent, Belgium*
[2] *Helsinki Institute of Life Science, Institute of Biotechnology, University of Helsinki, Finland*

4.1   Introduction

In life science research many technologies, like genomics, proteomics, metabolomics, and others, are applied to gather different information that contributes to the final understanding of how living organisms and cellular processes function. Microscopy is the technology that is used to obtain visual information and to show us the organization of a biological sample, which occurs at different levels, ranging from the cellular organization of a tissue to the subcellular patterning of organelles and cytoskeletal elements. Different microscopy techniques allow scientists to gather information at all these levels, varying from mm to µm, nm and even Å, to help us fully understand biological processes.

Different microscopic technologies will gather different kinds of information, which explains why a broad plethora of imaging techniques exists and continues to expand. Most commonly used techniques for life science research include: bright field microscopy, fluorescence widefield and confocal imaging, super-resolution techniques, and electron microscopy. Besides the variation in resolution, imaging technologies are different in the kind or content of information that is captured. Although the range of microscopic techniques is very broad, one can make a simple and clear-cut separation between light microscopy (LM) and electron microscopy (EM). This refers to the nature of the captured information in the final image, which comes from photons or electrons, respectively. While for LM the resolution limit is roughly 250 nm, EM can resolve structures in the nm and Å range.

LM is typically used for descriptive or functional information, depending on what dye or marker is visualized, and whether imaging is done on living or fixed specimens. Fluorescent dyes can be used with very high specificity to mark an object (organelle, protein of interest) in living or fixed cells or tissue. Major developments in this area came from developments in fluorescent molecules. For instance, the discovery of fluorescent proteins has opened up the possibility to genetically tag a protein of interest with a fluorophore and enables live cell imaging to capture functional information from living specimens (Giepmans et al. 2006). In addition, a broad range of LM techniques have been developed for different applications and make it possible to gain specific types of information, such as fluorescence resonance energy transfer to detect molecular proximity (Hirata and Kiyokawa 2016). However, apart from fluorescently tagged molecules, the rest of the cellular structure is invisible.

For a comprehensive view, EM is used because the power of EM lies not only in its resolution but also in the rich information content of the images. The addition of heavy metals or other electron dense compounds to a sample allows for the staining of lipids and proteins, resulting in the visualization of the ultrastructure of all organelles. The final EM image shows a very comprehensive view on what a sample looks like and it is not restricted to specific objects but rather the object of interest is viewed in its natural surroundings.

4.2   Imaging in 3D

There are constant technological developments in the field of microscopy, and some have responded to the need to collect 3D images, coming from the fact that biology happens in 3D. Before 3D imaging, in LM the thickness of the whole specimen was imaged in one dimension, including both in and out of focus information, resulting in reduced resolution. 3D LM became possible with the development of confocal fluorescence microscopes, which collect light emitted from the focal plane only (Cremer and Cremer 1978; Davidovits and Egger 1971; Sheppard and Kompfner 1978; Wilson 1989). By sequential imaging of different focal planes, a stack of images can be reconstructed to generate a 3D image. 3D LM is not destructive for the sample because light can travel through tissue and the focal plane can reside up to tens or a few hundreds of μm beneath the sample surface.

In transmission electron microscopy (TEM), a projection image from one thin section of the specimen is collected, while scanning electron microscopy (SEM) generates a topography image of the sample surface. Both technologies have been used to predict ultimate 3D structure, based on interpretation and extrapolation of the 2D images until different techniques were developed to collect 3D EM images (Peddie and Collinson 2014). One approach for 3D EM is the use of TEM tomography, where a 100–300 nm thick section is imaged from different angles and a volume image is reconstructed based on the series of collected projection images (Barcena and Koster 2009).

Another approach is to collect serial thin sections of the sample and image each section one after another by TEM or SEM (Micheva and Smith 2007; Tapia et al. 2012; Soto et al. 1994; Harris et al. 2006; Jokitalo et al. 2001). Manual sectioning is very labor-intensive, and even when sectioning is assisted with an automated tape collection device, it remains laborious (Hayworth et al. 2014). An alternative way to automate sectioning is the use of serial blockface SEM. Here, instead of imaging of sections, the smoothed block surface is imaged by SEM using a sensitive backscatter detector, and the next blockface is generated automatically either using a diamond knife (SBF-SEM) or focused ion beam (FIB) milling (FIB-SEM) (Denk and Horstmann 2004; Leighton 1981; Ishitani et al. 1995; Brkic et al. 2015; Vanwalleghem et al. 2015; Kremer et al. 2015). With this slice-and-view technique, the sample is eventually destroyed but the generation of stacks of EM images happens in relatively short time and without active input from the operator.

4.3   Comparative and Correlative LM and EM Imaging

A thorough analysis of a sample can only be achieved by complementation and combination of the various microscopy technologies. Very often, it is beneficial to link functional data from LM with comprehensive morphological information from EM. One can do this by preparing duplicate samples from the same specimen and imaging them side-by-side with at least two imaging modalities. In that case, two types of information are compared and each will aid to answer the research question. We refer to this approach as comparative light and electron microscopy (comparative-LEM).

A huge advantage of this method is that for each of the acquisition modalities the best sample preparation conditions can be followed. This is not the case with correlative light and electron microscopy (CLEM), where the exact same sample and region of interest (ROI) will be imaged with both LM and EM, and one single sample preparation protocol is applied. Since the optimal preparation methods can be very different for the two modalities, a compromise will have to be found that allows good imaging of the sample in both types of microscopes. Additional challenges have to do with the ability to find back the ROI in the two modalities, which happens at two stages in the process: (i) physical identification of the ROI in the sample when it has been transferred from LM to EM and (ii) digital identification of points within the ROI in the acquired LM and EM datasets.

4.4   CLEM Is More than LM + EM

Although CLEM is more demanding than comparative-LEM, the need for CLEM comes from the need to localize a specific point within an EM dataset. This can be the localization of a protein in its subcellular context, or a specific cell within a cell culture or tissue. In CLEM, the LM image is a tool to indicate where the needle in the haystack is residing and the EM image provides structural information at high resolution. For instance, fluorescent labeling of organelles such as mitochondria or endoplasmic reticulum (ER) will show where these organelles are localized, but EM images are required to determine whether there are changes in their structure and how they connect to other subcellular structures.

The final result is structural information from EM, but the approach of including LM into the workflow is necessary because of certain limitations of EM imaging. One is the lack of color contrast in an electron micrograph. Although there are tools that allow certain labels to stand out (e.g. HRP, APEX, photoconversion) as more electron dense and 'add color to EM' (Kremer et al. 2016; Ellisman et al. 2012; Puhka et al. 2007), generally all EM information is captured in grayscale values without any specificity in the signal intensity. This makes localization difficult, and therefore LM helps to put things in perspective and light up the needle in the EM haystack. Correlation with LM overcomes a second limitation of the EM, namely the very limited field of view that makes finding rare events challenging (Hellstrom et al. 2015; Spuul et al. 2011). With LM, one can image large numbers of cells very efficiently, or follow a distinct population of cells over time, to pinpoint the cell or area of interest for EM imaging. A third limitation of EM is that it cannot be applied to living cells. Correlation between live cell imaging and EM enables appreciation of dynamics and morphology from the same specimen.

4.5   3D CLEM

CLEM exists in many combinations, because for both LM and EM different techniques can be used. Initially, CLEM was carried out by correlating LM with 2D TEM or SEM (Yu et al. 2003), but over the past years volume EM has been used successfully in life science research, and this technology has also been part of CLEM workflows (Armer et al. 2009; Maco et al. 2013; Urwyler et al. 2015; Lees et al. 2017). Performing CLEM in 3D is important for the same reason any imaging technique is valuable in 3D: it gives a more correct, precise, and complete view. The incentive to perform 3D CLEM (with volume EM) as opposed to 2D CLEM is mainly out of scientific considerations. Volume EM is the defining part of a 3D CLEM project, and it can be following either 2D (*2D – 3D CLEM*) or 3D LM imaging (*3D – 3D CLEM*), and furthermore, the final result can be the volume EM dataset alone (from the area specified by LM), or in the most advanced cases, the integrated 3D LM and EM volumes from the same ROI (see Figure 4.1).

The practicalities that need to be worked out for any CLEM workflow, either 2D or 3D, boil down to (i) optimizing the sample preparation protocol, and (ii) introducing tools that allow to find back the ROI between the LM and EM acquisition steps. In 3D – 3D CLEM projects, additional complexity is introduced because of the volume of the sample, the resolution gap between 3D LM and volume EM imaging techniques, and the requirement to register exact coordinates for ROI in the both datasets to facilitate the final integration.

The sample preparation gap is due to the difference in imaging conditions between LM and EM. For LM, samples are typically imaged in conditions that are close to the normal habitat of the specimen with contrasting methods that interfere minimally with the sample. On the contrary, for EM the sample is subjected to heavy fixation and post-fixation followed by dehydration and plastic embedding because image acquisition occurs under vacuum. Live-cell imaging and intra-vital imaging are the most radical methods when it comes to performing microscopy on a specimen in its natural environments. They are done in well-controlled temperature and carbon dioxide conditions,

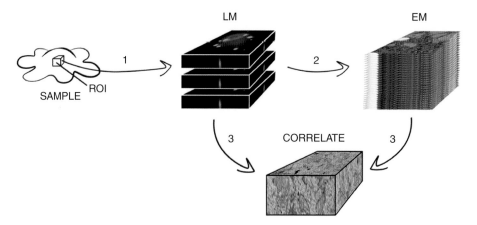

Figure 4.1 Schematic representation of a 3D CLEM project. Workflow starts with 2D or 3D LM imaging (1), which is followed by specimen preparation for EM and collection of a volume EM dataset from the corresponding ROI (2). The result can be only the volume EM dataset from the area specified by LM, or the two datasets can be integrated into a one-volume dataset (3).

the very opposite of fixation conditions that are essential for EM. Even when LM is performed on fixed samples, already from the fixation point the protocols for LM and EM separate as the most commonly used chemical fixative for EM, glutaraldehyde, cannot be used for LM because it is autofluorescent. In CLEM workflows, it is essential to minimize damage to the sample prior to the point at which the sample is fixed and prepared for EM. The condition of the cells should not be compromised before fixation and any light-induced phototoxic effects on the specimens have to be minimized. As a consequence, this can result in a compromise to the LM imaging conditions. Therefore, CLEM is only recommended for those projects where comparative-LEM is not sufficient to solve the biological question.

The next hurdle to take in 3D CLEM is to find back the ROI for the second image acquisition step. A major difference in acquiring a 3D LM dataset versus a 3D EM dataset is the fact that the LM image can be obtained from a certain depth within the specimen. For volume EM imaging, the very first image comes from the surface of the sample. So on the practical side, 3D CLEM requires the introduction of fiducial markers, either as part of the sample or artificial, to determine the edge of the ROI that should be prepared as the surface of the sample in EM.

When making use of cells grown on cover slips, the basal side of the cells will be immediately at the sample surface. However, when imaging tissue samples, the sample block needs to be trimmed to reach the ROI. It is essential to secure that during trimming no part of the ROI is lost. To be able to do that, the location of the ROI within the sample volume has to be known or determined. The efficiency of a 3D CLEM workflow, making use of SBF-SEM, will be determined by how quickly the ROI can be located and prepared as the sample surface. Also, the information outside the ROI could be part of the dataset as the slice and view technique can be carried out over a substantial depth (e.g. 2000 slices of 50 nm is realistic), however it is much more efficient to have the ROI as close as possible to the surface at the beginning of imaging and work with imaging parameters at optimal xy resolution for the field of view that covers the ROI. When making use of FIB-SEM for volume EM, it is even more crucial to have the ROI at the surface of the sample block, because this technique is more limited in the depth that will be imaged (about 20 µm).

Once both volume datasets are acquired, these have to be combined. For this step, the resolution gap between LM and EM as microscopy techniques adds complexity to this process. In 2D-3D CLEM, a single LM image corresponds to the whole 3D EM dataset, and functions mainly in specifying the cell, area, or organelle of interest. In 3D-3D CLEM, an LM-EM overlay can be performed quite accurately at x and y, but since the LM has even lower resolution in z, one LM image layer corresponds to several EM image layers (see Figure 4.1). For example, one layer from a confocal stack is typically around 500 nm thick, and corresponds to 10–20 blockface images taken at 30–50 nm intervals. Therefore, although in 3D CLEM all the Z-information is present in both datasets, fiducial markers at different z-levels are needed to provide the best fit for the overlay.

4.6   Two Workflows for 3D CLEM

In Figures 4.2 and 4.3, we provide two examples of 3D CLEM. The first example is a typical cell biology experiment, where the aim was to study the morphological alterations induced by silencing of a protein of unknown function. CLEM was needed for

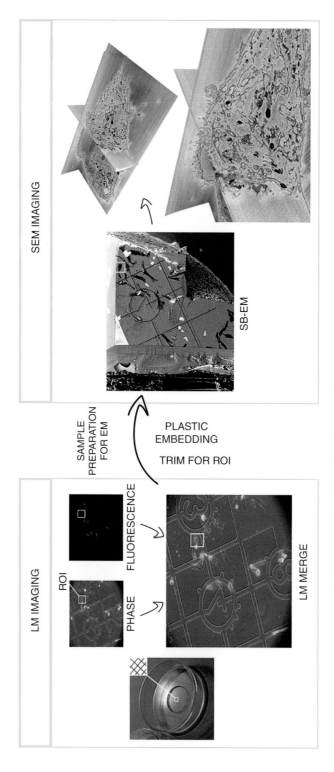

Figure 4.2 Schematic representation of a 3D CLEM workflow for adherent cells.

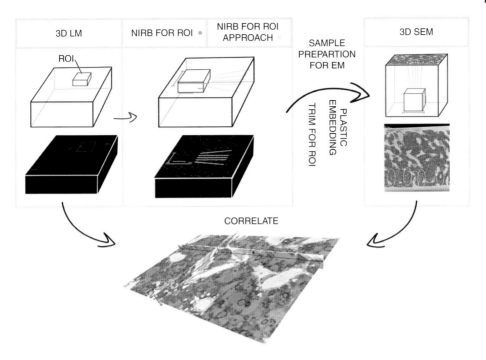

Figure 4.3 Schematic representation of a 3D CLEM workflow for tissue samples, making use of NIRB.

reliable identification of cells that were transfected with the silencing plasmid and 3D-EM to understand the alterations on organelle morphology.

Cells were grown on special glass coverslips that contain an etched grid on the surface, and transfected with two plasmids simultaneously. One plasmid is for silencing of the protein of interest and the other one is expressing fluorescent histone component (H2B-RFP) to identify transfected cells. Here, the transfection marker was chosen so that the fluorescence or expression of fusion protein do not interfere with the organelle of interest. We have earlier shown that when two plasmids are mixed together in 1:3 or 1:4 ratio (where smaller portion is the marker plasmid) and transfected simultaneously, the likelihood of cells being positive for both markers is over 95% (Joensuu et al. 2014). Cells were fixed with formaldehyde prior to phase contrast and fluorescent LM imaging. Phase contrast images revealed cells and the etched grid on the coverslip, and when overlaid with the fluorescent images, a map for localization of transfected cells was generated. After LM imaging, cells were processed for EM using flat-embedding protocol (Puhka et al. 2012; Seemann et al. 2000). During plastic embedding the imprint of grid on the coverslip was transferred onto the epoxy block. Next, a pyramid was carefully trimmed so that the selected cell of interest was located roughly in the middle of the trimmed area. Prior to serial blockface imaging, an image was taken with a high enough voltage that the beam penetrated the platinum coating to reveal both cells and etched markings. Then, serial blockface imaging from the corresponding ROI was done to generate a 3D volume of the transfected cell.

The second example comes from tissue where the ultrastructure of Kupfer cells, a macrophage cell type residing in liver tissue, needs to be examined. For this experiment a mouse model was used in which the Kupfer cells express tdTomato under the control of a cell specific promoter.

Mice were perfusion-fixed with buffered formaldehyde, the liver was dissected and sliced with a vibratome. Vibratome sections were used to collect 3D confocal images. Immediate processing of the sample for 3D EM data acquisition of the ROI was not possible without introducing a step in the workflow that helps to find back the ROI before starting to image the block surface. Sample preparation for volume EM requires *en bloc* staining, which turns the sample entirely black and opaque so that the ROI is hidden inside the sample block. It is necessary to remove sample tissue around the ROI so that at least one plane of the ROI is positioned at the surface. In order to determine when the ROI is reached during trimming of the sample, we used an approach that was previously published to perform 3D CLEM for brain tissue (Maco et al. 2013; Urwyler et al. 2015), making use of near infrared branding (NIRB) to introduce a scarring pattern in the tissue around the ROI (Bishop et al. 2011). Aternatively to NIRB, a 2-photon laser setup at high energy could be used for the same purpose. These marks were visible in both the LM and EM. The position of the marks in relation to the sample surface was measured in LM and once the tissue is *en bloc* stained and embedded for SB-EM, the area around the marks was trimmed away. The trimming requires carefulness and precision, to make sure that during this step the ROI would not be trimmed away. First, SBF-SEM was used with acquisition of images at lower magnification until the marks were visible in the datastack. Once the ROI was reached, the acquisition was performed at higher resolution. Since the NIRB marks were visible in both LM and EM, they were used as localization references between the LM stack and EM stack, too. This allowed us to find back the cell of interest that was first identified by fluorescence, in the 3D EM dataset.

In both examples, CLEM was achieved by introducing references that are present in the sample and became part of both the 3D LM and the 3D EM dataset. In both examples, fluorescent imaging was used to identify the target, but 3D-EM was needed for appreciation of the true structure.

4.7 Where Is CLEM Going in the Future?

There are several success stories in which 3D CLEM has been performed. Nevertheless, these methods are not performed at a high throughput, because the whole process is very long and labor intensive. In the first example, the sample preparation after the LM imaging took two full working days and two overnight incubations, preparations for SBF-SEM imaging half a day, and collection of the EM dataset was done in about 6 hours.

In the second example, the process took even more time: LM imaging took half a day, preparation of the sample for volume EM 5 days, approaching the region of interest and EM acquisition 72 hours, and data reconstruction for overlays required 5 days. It is good to keep in mind that the target in CLEM projects is always very restricted, typically one cell or part of a cell, and for reliable results and statistical significance the process has to be repeated several times. Because of this, 3D CLEM is at the risk of remaining rather specialized technology performed in most advanced imaging projects only.

In addition to two imaging modalities, the CLEM workflow comprises different segments, such as compatible sample preparation, clear indication of the ROI in LM, sample orientation between imaging modalities, and reference points in LM and EM.

Improvements in all these segments can contribute to the progress of CLEM as a methodology as much as any development done for improving the actual imaging steps.

When designing a CLEM project, the first consideration should be the biology, not the imaging requirements. For correlation purposes, high fluorescence signal is usually beneficial, and researchers may be tempted to boost the signal by high overexpression. Overexpression of soluble markers may lead to mislocalization of the marker (e.g. leaking of luminal ER proteins to the cis-Golgi) or aggregation in the cytosol. Overexpression of membrane proteins, on the other hand, may alter the morphology of membrane bound organelles for example by inducing filopodia formation at the plasma membrane, invaginations to the nuclear envelope or changing the sheet – tubule ration of the ER. Therefore, any development that increases the sensitivity for LM imaging is an advantage, because it allows imaging of fluorescent markers closer to endogenous levels. Moreover, for CLEM it is essential that the LM imaging technique has as few effects as possible on the general condition of the sample. For example, more sensitive detectors for fluorescence microscopy will result in less light exposure, and hence less phototoxicity.

Having survived the initial LM imaging part, the next challenge is the sample preparation for EM. Here, sample preparation improvements to ensure best morphological preservation will certainly be of benefit for CLEM workflows. One immediate application is the use of high-pressure freezing as the initial fixation method, followed by freeze substitution for sample embedding. Fixation by freezing occurs faster than chemical fixation, thereby making temporal correlation more accurate. Processing at low temperatures using freeze substitution may reduce extraction of small soluble components and lipids, protect against osmotic stress, and reduce shrinkage during dehydration. Fluorescent imaging can already be done under liquid nitrogen temperatures, but more work needs to be done to find good substitution protocols that give high enough contrast that SB-EM imaging requires. For 3D CLEM with the use of FIB-SEM as EM technology it is absolutely essential to have the ROI at the surface. This can be challenging, but promising workflows have been used in which correlation from *in vivo* confocal imaging to FIB-SEM imaging was aided by introducing an intermediate imaging modality, more specifically microCT (Karreman et al. 2017).

There are two steps where the LM and EM are brought together: (i) tracking down the ROI in the sample after transferring the sample from LM to EM, and (ii) determining the position of the EM dataset in relation to the LM image. Both are challenging because of the difference in resolution between LM and EM, this resolution gap definitely needs to be overcome to make the CLEM process better. If the motivation for doing CLEM falls in finding the needle in the haystack challenge, improved z-resolution might not be that important. However, as the technology becomes more widely used, more challenging experimental setups are requested where the resolution of the correlation might become the limiting point. Examples from cell biology include finding the initial formation points of autophagosomes or lipid droplets from the ER, or exact membrane contact points between various organelles.

Most super-resolution techniques have impressively improved resolution in xy. If similar improvements can be achieved in z-resolution, that would be a substantial step forward for CLEM. Ideally, in the future we would be able to increase the z-resolution in LM and EM close to the resolution that we currently have at xy, so that we would be able to operate at high isotropic resolution. The smaller the difference in resolution between the datasets, the more accurate the overlay. However, as the EM dataset always contains more information, bigger voxel size at LM dataset is acceptable. And even in case that LM techniques would be developed that reach the same resolution as EM, CLEM would still make sense because both modalities give different information that will be combined.

Finally, we expect that technical developments will lead to larger volumes that can be covered with correlative imaging, which would be welcomed as this increases the possibilities for statistical analysis of increased number of events or targets.

Clearly, there is a need for improved data reconstruction techniques. While, the LM datasets can be assembled in 3D relatively fast and easy, volume EM data reconstruction is laborious (Belevich et al. 2016; Heiligenstein et al. 2017; Maco et al. 2014; Paul-Gilloteaux et al. 2017). Especially, segmentation of objects from these datasets requires still a lot of manual input. Besides that, making the overlay images is difficult to perform with great precision and is far from a process that is performed automatically. Introduction of markers that can be visualized in both modalities can be a general tool that will allow this. Attempts toward using nanoparticles and crystals as markers in z have been made. These markers could be introduced into the sample by allowing them to bind to the outside surface of the cell, or cells can actively endocytose them using phagosytosis. Here, one technical challenge is to find suitable markers that do not cause damage to the diamond knife used in SB-EM. Many nanoparticles currently under testing are so hard that they easily damage the knife edge. A different approach is used for tissue, where one could add a ruler that indicates the z-position. If the NIRB is used to brand diagonal lines in the approach area above the ROI, then the xy distance can give an indication of the z-position (e.g. cone-shaped markers ∧, positioned as such that the lines are branded in the xz and zy field of the SB EM dataset with the base close to the ROI and the point far from the ROI will be recognized in the xy image as two scars). The distance between these will increase upon ROI approaching.

3D CLEM is a difficult methodology, but the need to gather volume data will only increase. In life science research we have gigantic access to information on cells, tissues, model systems, and disease conditions. All of this information has been gathered using different technologies, but imaging will still be the technique that gives us spatial localization for biological events. The fact that we can perform that in 3D is a tremendous added value to the complete understanding of biology. For sure, 3D CLEM processes will become more efficient, especially if current bottlenecks can be overcome, by speeding up the reconstruction of the volume datasets and the generation of correlative overlays. So real advances for CLEM will not only reside in gathering more volume information, or improve the speed at which the methodology can be carried out, but mainly in techniques that add new information that helps in being more precise in the process of overlaying/correlating two different kinds of images of the exact same object.

Acknowledgments

The authors would like to thank Martin Guilliams, Johhny Bonnardel, Anneke Kremer, and Evelien Van Hamme (VIB, UGent) for the CLEM experiment with liver tissue and Darshan Kumar, Ilya Belevich, and Mervi Lindman (Institute of Biotechnology, University of Helsinki) for the CLEM experiment with cultured cell. We thank Anneke Kremer,

Evelien Van Hamme and Chris Guérin (VIB, Ghent), and Helena Vihinen, Taina Suntio, and Leena Meriläinen (Institute of Biotechnology, University of Helsinki) for reading the manuscript and Maro Järvinen (Helsinki; http://maro.fi/) for preparing the figures.

References

Armer, H. E., Mariggi, G., Png, K. M., Genoud, C., Monteith, A. G., Bushby, A. J., et al. 2009. Imaging transient blood vessel fusion events in zebrafish by correlative volume electron microscopy. *PLOS ONE*, 4, e7716.

Barcena, M. and Koster, A. J. 2009. Electron tomography in life science. *Seminars in cell and developmental biology*, 20, 920–930.

Belevich, I., Joensuu, M., Kumar, D., Vihinen, H. and Jokitalo, E. 2016. Microscopy Image Browser: A Platform for Segmentation and Analysis of Multidimensional Datasets. *PLOS biology*, 14, e1002340.

Bishop, D., Nikic, I., Brinkoetter, M., Knecht, S., Potz, S., Kerschensteiner, M. and Misgeld, T. 2011. Near-infrared branding efficiently correlates light and electron microscopy. *Nature methods*, 8, 568–570.

Brkic, M., Balusu, S., Van Wonterghem, E., Gorle, N., Benilova, I., Kremer, A., et al. 2015. Amyloid beta Oligomers Disrupt Blood-CSF Barrier Integrity by Activating Matrix Metalloproteinases. *The Journal of neuroscience: the official journal of the Society for Neuroscience*, 35, 12766–12778.

Cremer, C. and Cremer, T. 1978. Considerations on a laser-scanning-microscope with high resolution and depth of field. *Microscopica acta*, 81, 31–44.

Davidovits, P. and Egger, M. D. 1971. Scanning laser microscope for biological investigations. *Applied optics*, 10, 1615–1619.

Denk, W. and Horstmann, H. 2004. Serial block-face scanning electron microscopy to reconstruct three-dimensional tissue nanostructure. *PLoS biology*, 2, e329.

Ellisman, M. H., Deerinck, T. J., Shu, X. and Sosinsky, G. E. 2012. Picking faces out of a crowd: genetic labels for identification of proteins in correlated light and electron microscopy imaging. *Methods in cell biology*, 111, 139–155.

Giepmans, B. N., Adams, S. R., Ellisman, M. H. and Tsien, R. Y. 2006. The fluorescent toolbox for assessing protein location and function. *Science*, 312, 217–224.

Harris, K. M., Perry, E., Bourne, J., Feinberg, M., Ostroff, L. and Hurlburt, J. 2006. Uniform serial sectioning for transmission electron microscopy. *The Journal of neuroscience: Official Journal of the Society for Neuroscience*, 26, 12101–12103.

Hayworth, K. J., Morgan, J. L., Schalek, R., Berger, D. R., Hildebrand, D. G. and Lichtman, J. W. 2014. Imaging ATUM ultrathin section libraries with WaferMapper: a multi-scale approach to EM reconstruction of neural circuits. *Frontiers in neural circuits*, 8, 68.

Heiligenstein, X., Paul-Gilloteaux, P., Raposo, G. and Salamero, J. 2017. eC-CLEM: A multidimension, multimodel software to correlate intermodal images with a focus on light and electron microscopy. *Methods in cell biology*, 140, 335–352.

Hellstrom, K., Vihinen, H., Kallio, K., Jokitalo, E. and Ahola, T. 2015. Correlative light and electron microscopy enables viral replication studies at the ultrastructural level. *Methods*, 90, 49–56.

Hirata, E. and Kiyokawa, E. 2016. Future Perspective of Single-Molecule FRET Biosensors and Intravital FRET Microscopy. *Biophys J*, 111, 1103–1111.

Ishitani, T., Hirose, H. and Tsuboi, H. 1995. Focused-ion-beam digging of biological specimens. *J Electron Microsc (Tokyo)*, 44, 110–114.

Joensuu, M., Belevich, I., Ramo, O., Nevzorov, I., Vihinen, H., Puhka, M., et al. 2014. ER sheet persistence is coupled to myosin 1c-regulated dynamic actin filament arrays. *Molecular biology of the cell*, 25, 1111–1126.

Jokitalo, E., Cabrera-Poch, N., Warren, G., and Shima, D. T. 2001. Golgi clusters and vesicles mediate mitotic inheritance independently of the endoplasmic reticulum. *The Journal of cell biology*, 154, 317–330.

Karreman, M. A., Ruthensteiner, B., Mercier, L., Schieber, N. L., Solecki, G., Winkler, F., et al. 2017. Find your way with X-Ray: Using microCT to correlate in vivo imaging with 3D electron microscopy. *Methods in cell biology*, 140, 277–301.

Kremer, A., Lippens, S., Bartunkova, S., Asselbergh, B., Blanpain, C., Fendrych, M., et al. 2015. Developing 3D SEM in a broad biological context. *Journal of microscopy*, 259, 80–96.

Kremer, A., Lippens, S., and Guerin, C. J. 2016. Correlative Light and Electron Microscopy: Methods and Applications. *eLS*.

Lees, R. M., Peddie, C. J., Collinson, L. M., Ashby, M. C., and Verkade, P. 2017. Correlative two-photon and serial block face scanning electron microscopy in neuronal tissue using 3D near-infrared branding maps. *Method Cell Biol*, 140, 245–276.

Leighton, S. B. 1981. SEM images of block faces, cut by a miniature microtome within the SEM – a technical note. *Scan Electron Microsc*, 73–76.

Maco, B., Cantoni, M., Holtmaat, A., Kreshuk, A., Hamprecht, F. A. and Knott, G. W. 2014. Semiautomated correlative 3D electron microscopy of in vivo-imaged axons and dendrites. *Nature protocols*, 9, 1354–1366.

Maco, B., Holtmaat, A., Cantoni, M., Kreshuk, A., Straehle, C. N., Hamprecht, F. A. et al. 2013. Correlative in vivo 2 photon and focused ion beam scanning electron microscopy of cortical neurons. *PloS one*, 8, e57405.

Micheva, K. D. and Smith, S. J. 2007. Array tomography: a new tool for imaging the molecular architecture and ultrastructure of neural circuits. *Neuron*, 55, 25–36.

Paul-Gilloteaux, P., Heiligenstein, X., Belle, M., Domart, M. C., Larijani, B., Collinson, L., et al. 2017. eC-CLEM: flexible multidimensional registration software for correlative microscopies. *Nature methods*, 14, 102–103.

Peddie, C. J. and Collinson, L. M. 2014. Exploring the third dimension: volume electron microscopy comes of age. *Micron*, 61, 9–19.

Puhka, M., Joensuu, M., Vihinen, H., Belevich, I. and Jokitalo, E. 2012. Progressive sheet-to-tubule transformation is a general mechanism for endoplasmic reticulum partitioning in dividing mammalian cells. *Molecular biology of the cell*, 23, 2424–2432.

Puhka, M., Vihinen, H., Joensuu, M. and Jokitalo, E. 2007. Endoplasmic reticulum remains continuous and undergoes sheet-to-tubule transformation during cell division in mammalian cells. *The Journal of cell biology*, 179, 895–909.

Seemann, J., Jokitalo, E. J. and Warren, G. 2000. The role of the tethering proteins p115 and GM130 in transport through the Golgi apparatus in vivo. *Molecular biology of the cell*, 11, 635–645.

Sheppard, C. J. and Kompfner, R. 1978. Resonant scanning optical microscope. *Applied optics*, 17, 2879–2882.

Soto, G. E., Young, S. J., Martone, M. E., Deerinck, T. J., Lamont, S., Carragher, B. O., et al. 1994. Serial section electron tomography: a method for three-dimensional reconstruction of large structures. *Neuroimage*, 1, 230–243.

Spuul, P., Balistreri, G., Hellstrom, K., Golubtsov, A. V., Jokitalo, E. and Ahola, T. 2011. Assembly of alphavirus replication complexes from RNA and protein components in a novel trans-replication system in mammalian cells. *J Virol*, 85, 4739–4751.

Tapia, J. C., Kasthuri, N., Hayworth, K. J., Schalek, R., Lichtman, J. W., Smith, S. J. et al. 2012. High-contrast en bloc staining of neuronal tissue for field emission scanning electron microscopy. *Nature protocols*, 7, 193–206.

Urwyler, O., Izadifar, A., Dascenco, D., Petrovic, M., He, H., Ayaz, D., et al. 2015. Investigating CNS synaptogenesis at single-synapse resolution by combining reverse genetics with correlative light and electron microscopy. *Development*, 142, 394–405.

Vanwalleghem, G., Fontaine, F., Lecordier, L., Tebabi, P., Klewe, K., Nolan, D. P., et al. 2015. Coupling of lysosomal and mitochondrial membrane permeabilization in trypanolysis by APOL1. *Nat Commun*, 6, 8078.

Wilson, T. 1989. Three-dimensional imaging in confocal systems. *Journal of microscopy*, 153, 161–169.

Yu, L. Y., Korhonen, L., Martinez, R., Jokitalo, E., Chen, Y., Arumae, U. et al. 2003. Regulation of sympathetic neuron and neuroblastoma cell death by XIAP and its association with proteasomes in neural cells. *Mol Cell Neurosci*, 22, 308–318.

5

Can Correlative Microscopy Ever Be Easy? An Array Tomography Viewpoint

Irina Kolotuev[1] and Kristina D. Micheva[2]

[1] University of Lausanne, EM Facility, Switzerland
[2] Stanford University School of Medicine, California, United States

5.1 Introduction

Array tomography is conceptually very simple. A three-dimensional biological sample is reduced to a two-dimensional array of ultrathin sections, which are easily accessible to immunostaining and imaging at both light and electron microscopic level. The obtained image stacks are then reassembled into a three-dimensional representation of the sample. From a practical point of view, array tomography combines electron microscopy methods of sample preparation with conventional immunofluorescence, and then applies image registration and volume rendering software tools to visualize the results. Thus, at least in theory, array tomography can be performed in any standard EM lab with access to a fluorescence microscope and a scanning electron microscope.

In practice, of course, there are numerous details at every step that may require adjustment, critical evaluation and trouble shooting for even the most experienced users. Avoiding folds in the sections, finding the right antibodies, and ensuring sufficient contrast for SEM imaging are only a few of the many concerns that can complicate a typical straightforward experiment. And such problems grow exponentially when scaling up to image larger volumes with more antibody stains at a higher resolution. Can correlative array tomography ever be easy? In this chapter, we explore different strategies for designing array tomography experiments with the aim of making this technique more easily accessible to new users with limited resources.

5.2 Why Array Tomography?

Array tomography (AT) is based on the planar arraying, staining and imaging of ultrathin serial sections [1], [2]. It can be used primarily for immunofluorescence detection of multiple markers, or for electron microscopic ultrastructural analysis, or in both modalities as conjugate light-electron microscopy. Iterative cycles of immunostaining, imaging, and antibody elution enable acquisition of dozens of immunofluorescence

Correlative Imaging: Focusing on the Future, First Edition. Edited by Paul Verkade and Lucy Collinson.
© 2020 John Wiley & Sons Ltd. Published 2020 by John Wiley & Sons Ltd.

channels, which can be computationally registered with the subsequently acquired SEM images.

The AT approach offers several important advantages for correlative microscopy. First, it can provide very-high-content information at the light level, including multiple antibody immunolabels (more than 10), additional markers like endogenously expressed fluorescent proteins, as well as very high resolution (typically $100 \times 100 \times 70$ nm voxels) and large volumes. Second, it enables robust registration between the light and electron microscopy channels because the same array of ultrathin sections is imaged with the two modalities. This is the reason why we refer to this approach as conjugate AT, which is a special case of stringent correlative microscopy that uses the exact same physical sample. And third, AT as a nondestructive technique permits a certain level of flexibility in experimental design. Additional rounds of immunostaining can be added, for example, in case of immunostaining problems, or if the already applied antibodies reveal unexpected features that require a follow-up. Subsequently, upon completion of the electron microscopic imaging, the samples can be revisited again at a later time and reimaged at the SEM at a different magnification, or in a different region. After all the data have been acquired and analyzed for a particular study, image stacks on interest can be examined with new questions in mind.

5.3 Array Tomography of Abundant Subcellular Structures: Synapses

AT was first developed for the study of mammalian central nervous system synapses. Because synapses are small submicron structures, densely packed in the brain, and diverse in their molecular composition, they are not easily accessible for a comprehensive study with other imaging methods. AT is particularly well suited for this subject of investigation, as it can fulfill the simultaneous need for multiplexed immunolabeling, individual synapse detection and large volumes. The small size and high density of synapses also make the AT approach straightforward, because there is no need to find a particular rare region of interest, as is the case with many other biological questions, a scenario that will be discussed later in this chapter. AT can be used to address a wide variety of questions, from the focused investigation of the distribution of a specific protein at specific synapses, to much broader questions such as: What are the different synapses composing a particular brain region [4]? How do synapses change during plasticity or disease [5], [6], [7]? Are there common patterns of synaptic connectivity [8], [9]?

Antibody labels will vary depending on the question, but it is always useful to include common synaptic markers, such as synapsin, synaptophysin, as well as markers that differentiate between excitatory and inhibitory synapses (Figure 5.1). Excitatory synapses can be distinguished by the expression of vesicular glutamate transporters (VGluTs) on the presynaptic side and the scaffolding protein PSD95 on the postsynaptic side, while inhibitory GABA synapses contain the vesicular GABA transporter (VGAT) and glutamic acid decarboxylase (GAD) on the presynaptic side, and gephyrin on the postsynaptic side [2]. Conjugate AT data with such general synaptic markers have shown that many synapses can be reliably detected by using only light microscopy [2], [3]. Synapse detection algorithms developed specifically for AT IF data achieve around

Box 5.1   A typical conjugate AT experiment for tissue samples [3].

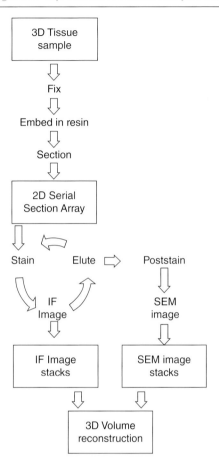

The sample is chemically fixed, usually through transcardial perfusion of the animal, or through immersion in fixative. Good ultrastructural preservation is achieved using a combination of formaldehyde and glutaraldehyde. The fixed tissue is cut into 100–200 μm slices using a Vibratome, and the region of interest is dissected out. Tissue dehydration is performed through freeze-substitution, after which the sample is infiltrated with the resin Lowicryl HM20 and polymerized using UV light at 0°C. Ultrathin serial sections are cut following standard electron microscopy procedures, with a diamond knife on an ultramicrotome. Usually 50–100 serial sections are picked on a carbon-coated coverslip. The sections are labeled using conventional indirect immunostaining, with up to three different primary antibodies, followed by the corresponding secondary antibodies conjugated to fluorophores with distinct excitation/emission spectra. The coverslips are mounted on glass slides or custom-made chambers using nonhardening mounting medium that also contains the nuclear stain DAPI, and are imaged using an automated fluorescence microscope. First, the coverslip is imaged at low magnification, typically with a 10× objective, to acquire a tiled image from the entire DAPI-stained ribbon. The desired region of interest is chosen on a section from the low-magnification map, and software tools are used to find the position of the same region on each serial section; the resulting list of positions is then imaged in an automated way at high magnification (63× objective). After imaging, the antibodies are eluted from the sections using a high pH solution, and a new combination of antibodies is applied and imaged. This cycle can be repeated many times. After completion of the immunofluorescent imaging, the antibodies are eluted again, the ribbon is washed well and poststained with heavy metals for electron microscopic imaging. Ribbons are imaged in an FESEM microscope using the backscattered electron detector, typically focusing on a small subvolume from the region imaged with immunofluorescence. Finally, IF and SEM images are registered to assign the immunofluorescence from different channels to the underlying ultrastructure.

90% accuracy for excitatory synapses in rodent cortex, as verified by comparing with ultrastructurally defined synapses from conjugate SEM images [3], [10]. This points to one obvious way of simplifying conjugate AT and making it more efficient: by relying predominantly on immunofluorescence data, which can be obtained much more easily and rapidly compared to EM data. Small SEM subvolumes can be used to develop and confirm rules of IF detection of different synapse types, which are then applied to the large volumes with IF data. The same principle can be applied to other biological problems as well. For example, the myelinated sheaths of axons are reliably detected by using immunofluorescence for abundant myelin proteins, such as myelin basic protein (MBP) or proteolipid protein (PLP), as confirmed with conjugate AT [11], which enables large-scale studies of myelinated axons by predominantly IF AT.

5.4 Array Tomography of Sparsely Distributed Structures: Cisternal Organelle

Many interesting subcellular structures are much sparser compared to central nervous system synapses. While they span only several ultrathin sections and can be easily imaged at very high resolution with electron microscopy once identified, the problem here becomes in actually finding them. In such cases, conjugate AT can aid in locating

Figure 5.1 **Conjugate AT of synapses in mouse neocortex. a.** An overlay of the fluorescent nuclear label DAPI and the SEM image from a section through the mouse neocortex. **b.** Higher magnification view of the SEM image in A, overlaid with immunofluorescence for synapsin (general presynaptic marker, green), PSD95 (postsynaptic density protein 95, postsynaptic marker of excitatory synapses, red), GABA (gamma-aminobutyric acid, inhibitory neurotransmitter, cyan), and MBP (myelin basic protein, magenta). **c.** Left panel, higher-magnification view of the boxed region in B, with the same immunolabels. Right panels, additional markers on the numbered synapses in the left panel. GAD2 (glutamate decarboxylase 2, presynaptic marker for GABAergic inhibitory synapses, magenta), gephyrin (postsynaptic marker of inhibitory synapses, yellow), and VGluT1 (vesicular glutamate transporter 1, presynaptic marker of the majority of excitatory synapses in cortex). Image stacks courtesy of Forrest Collman, Michelle Naugle, and Stephen J Smith, Allen Institute for Brain Science.

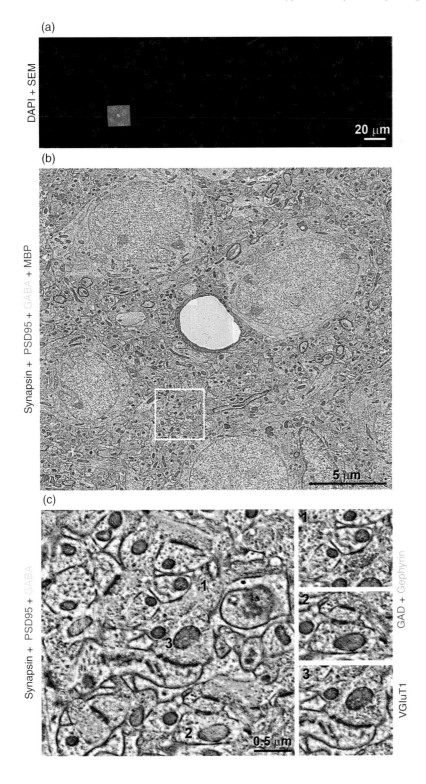

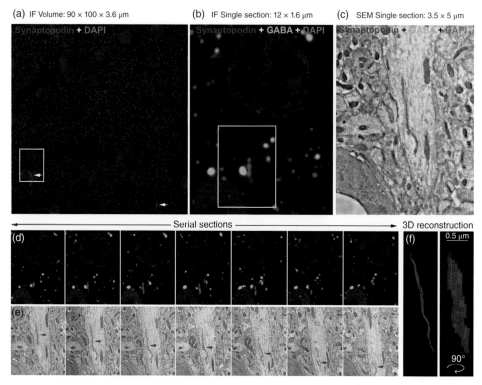

(a) IF Volume: 90 × 100 × 3.6 μm   **(b)** IF Single section: 12 × 1.6 μm   **(c)** SEM Single section: 3.5 × 5 μm

Synaptopodin + DAPI   Synaptopodin + GABA + DAPI   Synaptopodin + GABA + DAPI

Serial sections — 3D reconstruction

(d)   (f) 0.5 μm

(e)   90°

Figure 5.2 Cisternal organelles in mouse neocortex. a. Reconstructed volume from 52 serial ultrathin sections immunostained with synaptopodin (red). The rod-like synaptopodin-positive structures are cisternal organelles (arrows), and the punctate synaptopodin immunofluorescence is localized predominantly to dendritic spines where it labels the spine apparatus. The nuclear label DAPI is in blue. **b.** A single section view from the boxed region in A. **c.** SEM view from the boxed region in B, overlaid with immunofluorescence for synaptopodin (red) and GABA (green). The synaptopodin immunofluorescence is localized to the cisternal organelle in an axon initial segment, which receives a GABAergic synapse in close proximity to the cisternal organelle. **d.** Serial sections through the region shown in B. **e.** SEM view from the same serial sections as in D, showing the cisternal organelle. **f.** 3D reconstruction of the cisternal organelle from the SEM images.

rare events for subsequent imaging at the SEM. One such interesting structure is the cisternal organelle that is found in the axon initial segment of some neurons. The cisternal organelle resembles the spine apparatus in dendritic spines and is made up of stacks of endoplasmic reticulum [12], [13]. Like the spine apparatus, it has a high concentration of synaptopodin [14]. It is though to be involved in Ca^{2+} regulation [14], [15], [16], and may have a role in plasticity [17], but its precise function is unclear. Using synaptopodin immunofluorescence, the cisternal organelle can be identified as a brightly stained rod-shaped structure among a sea of much smaller puncta likely labeling dendritic spines (Figure 5.2). The same structure can be easily relocated at the SEM, using either its coordinates if the imaging system allows it (e.g., with the Zeiss Shuttle and Find, or FEI MAPS, or an integrated microscope), or simply by counting the sections and then following the pattern of nuclei and blood vessels for orientation within the section.

5.5   Array Tomography of Small Model Organisms: C. elegans

Small-model organisms like *C. elegans* or *Drosophila melanogaster* present different challenges compared to the mammalian brain. Because of their small size, it is difficult to impossible to dissect out specific organs before the preparation and it is difficult to localize them inside the organism after preparation. Additionally, unlike other more homogeneous samples where one can collect the sections in any orientation, in asymmetric samples the position and orientation of sectioning are important, not only to aid in locating the region of interest, but also to facilitate data analysis and interpretation. With asymmetric specimens, to reach statistical significance in the analysis, it is not enough to use alternative regions of the same sample or several images from different sampling zones, but individual carefully oriented samples have to be analyzed one by one. This operation significantly lengthens the time necessary for the overall analysis. However, unlike, for example, connectomics studies of the mammalian brain, the question regarding a structure of interest rarely requires more than 200–300 sections 60–100 nm thick. Thus, multicellular asymmetric samples fall between two stools: on one hand, it is too tedious to analyze such samples using serial sectioning TEM, and, on the other hand, the effort necessary to obtain the data using volume analysis methods such as serial block-face SEM and focused ion beam SEM is not always justified. AT provides an intermediate approach, allowing data acquisition in a faster and more efficient way.

We suggest several modifications to the original AT procedure to adapt it to small organisms. The first step involves sample preparation. In this case, flat embedding works best because it enables a clear view of the sample and facilitates its orientation for trimming and sectioning. Tight trimming of the blocks using a trimming diamond knife further delimits the region for EM analysis [18], [19] (Figure 5.3a). Preparing the block in this way helps to save time during sectioning, as it allows skipping semi-thin sectioning, which is otherwise required to obtain a clear view of the sample. The next step is the actual generation of the arrays and their transfer to the support. Different solutions already exist to facilitate this, and the choice will greatly depend on the number and the size of the sections that need to fit on a given support. In many cases, it is helpful to mount several ribbons of serial sections on the same support. An example of a sectioning and transfer solution is presented in Figures 5.3b and 5.3c; [48]. It includes a specially designed diamond knife equipped with an enlarged boat and a bottom draining system. An entire glass coverslip or a segment of a wafer fits inside the knife boat. After the desired number of ribbons is generated, the water is drained from the knife boat and the ribbons are easily positioned in sequential order on the support.

Finding the specific region of interest (ROI) to be imaged on the mounted ribbons can be a very tedious process due to the large number of sections. However, unlike the previous examples with brain tissue samples, where it is usually necessary to use antibody markers or other fluorescent labels to identify structures, in model organisms the anatomy itself and the ultrastructure of the particular cells can guide the search for the ROI. Based on this feature, we developed a strategy that facilitates data collection and analysis without the use of additional markers.

In this *leaping* strategy, the section arrays are rapidly screened to discriminate between relevant and irrelevant sections (Figure 5.3d,e). The ribbons of consecutive sections in the array are collected serially and can be aligned alongside each other as

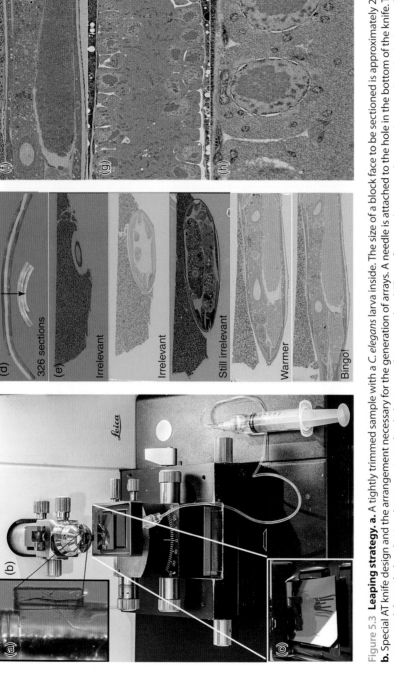

Figure 5.3 Leaping strategy. a. A tightly trimmed sample with a *C. elegans* larva inside. The size of a block face to be sectioned is approximately 250 × 150 µm. **b.** Special AT knife design and the arrangement necessary for the generation of arrays. A needle is attached to the hole in the bottom of the knife. This needle is connected through the plastic tube to a syringe that helps to regulate water level. The wafer or a glass coverslip are submerged inside the basin of the knife and covered with water. When the required amount of sections is generated, the water can be drained, while sections arranged on the support. **c.** The knife boat with a half-dried wafer revealing the sections on the dry surface. It is important to let the sections dry slowly in order to avoid micro folds on the surface of the sample. **d.** The aligned ribbons of sections arranged on a wafer. Overall, about 350 sections were collected serially. Each ribbon is pseudo-colored to highlight the relative position of the sections during the lateral screening. **e.** With the leaping strategy, the central section in each ribbon is sampled, and its proximity to the ROI is assessed based on anatomical features. **f.** A particular region of the gonad of *C. elegans* adult larva (highlighted in yellow). When the general region of interest is recognized, one can start acquiring images of a higher quality. Using this particular section as an anchor/point 0, serial acquisitions of each region of interest might be generated collecting the data from the sections upstream and downstream of this reference point. Scale bar, 10 µm. **g.** A zoom on the region of interest inside the gonad. Scale bar, 10 µm. **h.** A further zoom on a particular cellular event localized after the analysis of the image in panel G. Scale bar, 1 µm.

explained above. The length of each ribbon can be decided based on the expected frequency of the feature of interest. This way, instead of starting from one end of the ribbon and analyzing sections one by one, leaping from one ribbon to the next significantly speeds up the search for the ROI. Neighboring ribbons of sections are analyzed horizontally starting from a random section at the center of each ribbon. Once the ribbon that contains the ROI is identified, the sections are imaged serially, starting from the "reference anchor" section and going in both directions for as long as it is judged relevant. Because the sections remain on the support surface, they can be revisited using different acquisition parameters and sampling areas. This strategy can be used not only for purely ultrastructural investigations but also for correlative studies that incorporate appropriate fluorescent markers, with the leaping screen performed at the light level.

Small-model organisms such as *C. elegans* or *Drosophila melanogaster* offer the advantage of fast and cost-effective genetic manipulation, which has resulted in the generation of a multitude of strains with fluorescent markers targeted to specific cellular populations. As a result, thanks to the different publicly available constructs and strains, one can quickly identify the cells of interest and localize cell structures in a live organism. Fluorescence can also guide the localization of ROI within the resin block and on the array of sections, provided the fluorescence is preserved after embedding in resin, following established protocols [20], [21], [22]. This would be much harder or impossible to achieve on TEM grids due to their small size, while the AT approach can accommodate much larger samples and long ribbons of sections on the same substrate. The strategy of using endogenous fluorescent markers not only facilitates the identification of the structure of interest, but it also allows correlating the dynamics of the structure as observed in the live state with its ultrastructure.

We applied this strategy to image the mesenchymal distal tip cell of *C. elegans* gonad. The distal tip cell provides the niche for germline stem cells, and is often studied in relation to gonad migration, cell signaling, and cell–cell interactions [23], [24]. To visualize the interactions of the distal tip cell and the gonad during development and compare it among different mutants using EM, it is important to generate the longitudinal images that will provide insight on both types of cells. However, after high-pressure freezing and freeze substitution, animals rarely remain straight, and thus it is complicated to orient the plane of interest in parallel to the body axis. Random sectioning of the samples with further assessment of the fluorescence provides an excellent means for the selection of the ROI in a correct orientation.

In case of successful fluorescence preservation, the fluorescence remains visible not only inside the blocks but on sections as well (Figure 5.4a, b), eliminating the need for further labeling of the sample. In this manner, after sectioning, a simple screening can reveal the ROIs directly, and a subsequent EM analysis can be started right away (Figure 5.4b). After finding the cell or structure of interest, one can continue with the acquisition of the necessary volume of sections and either superimpose this information with the fluorescent data or just organize it in a 3D stack (Figure 5.4c). Cytonemes, the extensions essential for the proper signaling, are protruding from the distal cell body and cap the leading portion of the gonad (Figure 5.4c, green arrow). Tracking these structures on a single section or without a fluorescence reference can be a difficult task, as sometimes one part of the cytoneme is not attached to the rest of the extending structure. If required, sections can also be labeled with antibodies, before the EM observations, as the GFP signal remains detectable even after the additional labeling procedure, which can further facilitate the ultrastructure recognition process.

(a)   (b)

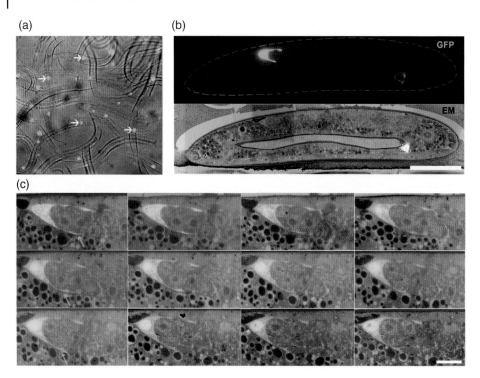

(c)

Figure 5.4 **AT localization and 3D CLEM of *C. elegans* gonad distal tip cell. a.** *C. elegans* larvae after HPF-quick freeze substitution embedded in HM20 resin. Transmission light and GFP fluorescence overlaid. Arrows point on gonad distal tip cell (DTC). **b.** A single tangential section through *C. elegans* adult larva body shows the fluorescence channel and an inverted contrast SEM image acquired with the backscattered electron (BSE) detector. The GFP signal, in this case, helps to localize the relevant sections rapidly among the numerous sections present on the wafer. Scale bar, 50μm. **c.** An overlay of the GFP fluorescent channel and the Back Scattered Electron SEM of serially acquired distal tip images. Arrows point to the cytonemal extensions of the DTC, easily visible by fluorescence, yet less evident when only the EM data is analyzed. Scale bar, 10μm.

5.6   To Summarize: Finding the Right AT Approach

As these examples show, there are many ways to use correlative AT. Different types of questions require different approaches, with varying levels of sophistication and automation. While some need highly specialized equipment, software and technical skills, others are better addressed using readily available and established approaches. For example, imaging of relatively small volumes does not necessarily require automated solutions for sections generation (as the ATUMtome, [25]), or for region relocation. In such cases, sections can be cut on a conventional ultramicrotome and manually collected, and following immunofluorescence imaging, the same region of interest can be easily relocated in the SEM by using reference points like blood vessels, nuclei and myelin (Figure 5.1), or the body anatomy and ultrastructural landmarks (Figure 5.3). In many cases, small tweaks significantly reduce the amount of effort and data needed to answer a question, as for example the "leaping" screening strategy used for *C. elegans*. The same technique can be applied for other similar samples, such as *Drosophila*, zebrafish embryonic stages, or Arabidopsis root. In addition, software strategies developed and validated using conjugate AT (e.g. synapse detection algorithms [3], [10]), enable data analysis based predominantly on IF data, resulting in faster acquisition of larger volumes. The most efficient way to approach a correlative microscopy problem will also vary, depending on the available resources and technical skills of the particular research lab. Recently, a lot of effort has gone into making AT large-scale and automated [26], [27], which enables large-volume multichannel dense reconstructions as needed, e.g., for connectomics or synaptomics studies. Such efforts benefit the entire research community, both by improving the tools and standardizing the protocols, as well as by acquiring and providing open access to high-quality datasets that are very rich in content and can be used to address a variety of questions. On the other hand, scaling up requires automation, specialized instruments, software, and data storage, which can become a high barrier to adopting AT in a small lab or for trying it out for preliminary results. With this in mind, we find it crucial to invest in a parallel effort to provide a suite of simple and accessible AT approaches that can be executed while using the instruments already available in typical EM labs, as done in the examples presented above (Figures 5.2, 5.3, and 5.4), without the need for additional expensive equipment.

5.7 Areas of Improvement

In addition to choosing the optimal experimental strategy, a number of specific areas require improvements to increase the accessibility and efficiency of AT as well as many other correlative microscopy approaches. Here are some of them.

5.7.1 Resin

Currently, the two resins that are routinely used for AT are LRWhite and Lowicryl HM20. LRWhite is a hydrophilic acrylic resin with relatively low toxicity, which is very straightforward to use and gives excellent and reproducible results for immunofluorescence AT. Its ability to preserve good ultrastructure for electron microscopy, however, is more problematic, even though in some cases it has been successfully used for correlative microscopy [21], [22], [28]. And because LRWhite polymerization is inhibited by oxygen, it is less convenient to flat embed samples to preserve a specific orientation. A preferred resin for electron microscopy is Lowicryl HM20, a hydrophobic acrylic resin, which is more toxic than LRWhite, harder to use, and less reproducible, but when used with freeze-substitution, it results in excellent ultrastructural preservation and reasonable antigenicity [3], [29], [30], [31]. The preparation of freeze-substituted and Lowicryl embedded tissue, however, is not trivial and requires extensive practical experience. Ideally, a robust resin embedding protocol can be developed that combines the ease of use of LRWhite with the excellent EM performance of Lowicryl.

5.7.2 Serial Ultrathin Sectioning

Ultrathin sectioning is a standard but highly skilled technique used in electron microscopy. One of the most difficult tasks is the generation of long uninterrupted series of ultrathin sections and their transfer to a solid substrate (usually a glass coverslip or silicon wafer), as done for AT. Some AT experiments can be accomplished by using rather short ribbons of 20 to 50 sections, but others require hundreds and even thousands of uninterrupted serial sections. There is a very long list of potential problems that even the most experienced user will eventually encounter. Ribbons curve, break, or attach to a wrong part of the substrate. Mounted sections have folds or large amounts of debris. Sections don't attach well to the substrate and come off during staining or antibody elution. The myriad problems of serial ultrathin sectioning have motivated different adaptations aimed at making this process more reliable and user-friendly. For example, one of the critical and most delicate steps of the sectioning process, the transfer of the section ribbons to the substrate, can be assisted by using the modified diamond knife boat described above [Burel et al., in preparation], or a custom-built substrate holder [32], or even magnetic field to guide the sections onto the substrate [33]. Automation of the entire process, as done with the ATUMtome, which uses a tape to collect each section as it comes off the diamond knife, allows the collection of thousands of serial sections [25]. The tape with the sections can be mounted on silicon wafers and imaged in the SEM; however, it is not suitable for high-resolution fluorescence imaging, and therefore for conjugate AT. While automated collection of serial sections on glass substrates is not yet possible, these recent technological advances are significantly improving the process and allowing the collections of thousands of serial sections for correlative microscopy.

5.7.3   Antibodies

Antibodies are crucial for AT. However, most commercial antibodies are not validated for use on plastic-embedded sections, and finding the right antibody is often a long and expensive process. A number of controls have to be performed to ensure that an antibody is specific and sensitive enough to be used for AT and software tools are now being developed to aid in this process [10]. The availability of antibodies is even scarcer for small model organisms, like *C. elegans* or *Drosophila*, and in such cases major efforts are put into the generation of transgenic fluorescent animals. Information and reagent sharing is one solution that we will discuss in more detail below.

5.7.4   EM Compatible Fluorophores

A major challenge for sample preparation is the preservation of the fluorescence. While there are now established protocols to preserve GFP and YFP fluorescence [3], [20], [21], [22], other endogenously expressed fluorophores, such as TdTomato, are more difficult to preserve. Antibody staining on sections can help in these cases, but it is often not enough to fully stain thinner cellular processes. This can cause difficulties in certain applications, such as when tracing the thin axons of neurons. A variety of new EM compatible fluorophores have been introduced recently (e.g. [34]), including fluorophores that withstand osmium fixation [35], [36].

5.7.5   Detectors and EM Resolution

The resolution provided by SEM is still significantly below what is possible with a TEM. And while the current SEM resolution is in many cases sufficient, there are questions that require better detectors. For example, identifying gap junctions or following of very thin cellular processes across serial sections require higher resolution than what SEMs currently provide. One workaround that has been successfully applied uses TEM instead of SEM: after completion of the fluorescent imaging, the sections are detached from the hard substrate (e.g. coverslip) and mounted on a grid to be examined in the TEM [9], [37].

5.7.6 Image Registration and Alignment Tools

In theory, conjugate AT presents a relatively straightforward case for registration of light and electron images, because the exact same physical object is imaged in the two modalities and there is no 3D ambiguity. It is important to have good reference fluorescence channels that can be matched to ultrastructural features. For example, the DAPI stain precisely maps the electron-dense heterochromatin in nuclei, and myelin basic protein immunolabel outlines the myelin sheath of axons. Autofluorescence can be used to identify lipofuscin granules or blood vessels and match to the corresponding features in the electron micrograph. Even with good reference features in the two modalities, image registration can be very tedious in practice, as it involves large differences in the scale of images coming from the fluorescence microscope and the SEM. In addition, defects in the sections (e.g. debris, folds, or scratches) can further complicate this task. There is still a need for robust, flexible, and fast software tools to perform image registration and alignment.

5.7.7 Data Sharing

The data generated by correlative microscopy is very rich, including both molecular and ultrastructural information. As such, its value often extends beyond the original scope of a particular experiment and can be an important resource for the scientific community. Data sharing is, of course, also needed for the usual workflow of a collaborative study. However, correlative microscopy data is often huge in size, and can have a complicated structure that includes the raw image files and various transformations for image registration and alignment. One excellent solution for sharing of big neuroscience related data is the Neurodata project (neurodata.io), which not only provides cloud storage but also enables cloud computing to visualize, analyze, and model the data.

5.7.8 "Dream" Resource

Nowadays, more than ever, science is a communal undertaking. The ease of online communication, the proliferation of highly specialized methods, and the shortage of funding are some of the many factors that enable and motivate researchers to collaborate and share expertise, reagents, and data. Correlative microscopy is a field that already greatly benefits from collaborations. Here we make the case for expanding such interactions by creating an online peer-to-peer network to share AT resources such as tissue blocks, arrays, and antibody aliquots.

How would this work? Judging from our own experience, labs that use AT, or electron microscopy in general, have drawers and drawers full of samples, such as tissue embedded in resin. Typically, for every experiment researchers prepare much more tissue than needed, often planning for unexpected complications, or simply because it is just as easy to prepare 10 blocks of tissue as it is 1 block. Similarly, when sectioning for AT, one prepares more arrays than usually needed. These extra blocks or arrays can be easily shipped using standard mail and will be of immense help to a beginner who would like to try out the technique, or for someone who needs to produce just one good image of a favorite protein immunolocalization. For a more experienced AT user, such a resource offers the chance to easily explore different tissues, species, or fixation protocols. This resource could also bring relief to the usually costly search for the right antibodies, as it can allow users to sample small aliquots shared by others. A couple of microliters will often be enough to decide whether an antibody has any promise for this application. The AT sharing network can work on an exchange basis, by offering tissue blocks, arrays, antibody aliquots, and even fixed tissue from a variety of specimen including transgenic mouse lines, different species and so on. Such a network will make AT much easier for beginners, and will significantly lower the costs and time for all users. It will give researchers access to a vast variety of samples from different species, brain regions and fixation/embedding conditions.

5.7.9 Dream Experiments

Despite the focus on making correlative microscopy easier and more accessible, our 'dream' experiments are anything but easy, at least right now. We would like to be able to gather and process much larger amounts of data in a faster and more efficient way, and we would like to do it better – with more reliable molecular information, with higher resolution. We would also like to make correlative microscopy even more correlative, that is, to include other modalities, like live imaging [38], electrophysiology [39] and transcriptomics [40].

One of the inherent drawbacks of AT is that this technique provides a snapshot in time. It will benefit immensely from integrating live imaging into the experimental process, and from the ability to correlate dynamic information about the biological event under study with the subsequent molecular and ultrastructural analysis. Live imaging has provided deep insights into numerous developmental and cell biological processes, such as organogenesis, tissue remodeling, cell migration, tissue polarity establishment, cell division, and the abscission, lysosome formation, endocytosis, multivesicular bodies formation and trafficking, and much more. There comes a point, however, where further understanding of many of these processes is hampered because of the lack of sufficient resolution and the difficulty in obtaining detailed molecular information when using only live microscopy. An integrated approach of live imaging followed by conjugate AT would maximize the advantages of both methods. This requires the ability to detect and "freeze" a transient event as it is happening, which currently usually involves sophisticated preparation procedures such as high-pressure freezing [41], as well as an efficient strategy for relocating the live imaged region – for example, by using gridded substrates for monolayer cell cultures, or 3D laser branding in the case of tissue samples [42].

In addition to live imaging, data obtained with other modalities will further enrich and complement the AT analysis. An example of a dream experiment from the neuroscience realm would be a comprehensive study that begins with a behavior and live imaging (e.g., [43], [44]), followed by relocation of the same live imaged brain region for conjugate AT [38], [39], including dense reconstructions [45], as well as in situ hybridization [40]. One could even envision a similar type of experiment directly addressing human brain functioning. For example, during brain surgeries, human intracranial electrophysiology is often used to identify noncritical brain tissue that can be removed without significantly affecting the brain function of the patient [46], [47]. Implanted electrodes during such surgeries can record the activity of single neurons in the patient during the performance of specific tasks. Upon resection of the tissue and its processing for conjugate AT, the reconstruction of the recorded neurons would reveal their synaptic connections (input and output), as well as their molecular composition, allowing the direct correlation of function, structure, and connectivity at the level of individual neurons, and this could be further extended to pairs or even larger numbers of interconnected neurons.

To make these experiments feasible, AT has to become much more reliable and faster, and its integration with other modalities less cumbersome. One would need to be able to progress in a short amount of time, for example, one day, from a tissue block to the EM images of the volume that are correlated to its LM information. Considering the pace of the current technological improvements, it is not a far stretch to hope that in the next 5 to 10 years we will have an LM-EM integrated AT machine, the output of which will be multicolored micrographs with the 3D data rendering directly available in the cloud for viewing and analysis by multiple researchers around the world. The results of even one such comprehensive experiment would provide data that will contain clues to many different biological problems, and the dissemination of such data will ultimately enable the most accessible form of AT, open AT.

Acknowledgments

The authors are grateful to Drs. Stephen J Smith, Forrest Collman and Michelle Naugle from the Allen Institute for Brain Science for sharing the conjugate AT dataset used for Figure 1, to JoAnn Buchanan for the SEM images for Figure 2, and to Drs. Mihoko Kato (Caltech) for the *C.elegans* strains. We would also like to acknowledge the help of the members of the EM facility of the University of Rennes 1 and the EM facility of the University of Lausanne. This work was supported by NIH grants R01NS092474, R01NS094499, and R01MH111768.

References

1 Micheva KD, Smith SJ (2007) Array tomography: a new tool for imaging the molecular architecture and ultrastructure of neural circuits. *Neuron* **55**(1):25–36.

2 Micheva KD, Busse B, Weiler NC, O'Rourke N, Smith SJ (2010) Single-synapse analysis of a diverse synapse population: proteomic imaging methods and markers. *Neuron* **68**(4):639–53.

3 Collman F, Buchanan J, Phend KD, Micheva KD, Weinberg RJ, Smith SJ (2015) Mapping synapses by conjugate light-electron array tomography. *J Neurosci* **35**(14):5792–807.

4 Soiza-Reilly M, Commons KG. (2011) Quantitative analysis of glutamatergic innervation of the mouse dorsal raphe nucleus using array tomography. *J Comp Neurol.* **519**:3802–14.

5 Koffie RM, Hashimoto T, Tai HC, Kay KR, Serrano-Pozo A, Joyner D, Hou S, Kopeikina KJ, Frosch MP, Lee VM, Holtzman DM, Hyman BT, Spires-Jones TL. (2012) Apolipoprotein E4 effects in Alzheimer's disease are mediated by synaptotoxic oligomeric amyloid-β. *Brain* **135**:2155–68.

6 Rasakham K, Schmidt HD, Kay K, Huizenga MN, Calcagno N, Pierce RC, Spires-Jones TL, Sadri-Vakili G. (2014) Synapse density and dendritic complexity are reduced in the prefrontal cortex following seven days of forced abstinence from cocaine self-administration. *PLoS One.* **9**(7):e102524.

7 Wang GX, Smith SJ, Mourrain P. (2016) Sub-synaptic, multiplexed analysis of proteins reveals Fragile X related protein 2 is mislocalized in *Fmr1* KO synapses. *Elife 5.* **pii**: e20560.

8 Oberti D, Kirschmann MA, Hahnloser RH. (2010) Correlative microscopy of densely labeled projection neurons using neural tracers. *Front Neuroanat.* **4**:24.

9 Bloss EB, Cembrowski MS, Karsh B, Colonell J, Fetter RD, Spruston N. Structured Dendritic Inhibition Supports Branch-Selective Integration in CA1 Pyramidal Cells. *Neuron.* 2016 **89**(5):1016–30.

10 Simhal AK, Aguerrebere C, Collman F, Vogelstein JT, Micheva KD, Weinberg RJ, Smith SJ, Sapiro G. (2017) Probabilistic fluorescence-based synapse detection. *PLoS Comput Biol.* 2017 Apr 17;**13**(4):e1005493.

11 Micheva KD, Wolman D, Mensh BD, Pax E, Buchanan J, Smith SJ, Bock DD A large fraction of neocortical myelin ensheathes axons of local inhibitory neurons. *Elife* 2016 5. **pii**: e15784.

12 Peters A, Proskauer CC, Kaiserman-Abramof IR. The small pyramidal neuron of the rat cerebral cortex. The axon hillock and initial segment. *J Cell Biol.* 1968 **39**(3):604–19.

13 Kosaka T. The axon initial segment as a synaptic site: ultrastructure and synaptology of the initial segment of the pyramidal cell in the rat hippocampus (CA3 region). *J Neurocytol.* 1980 **9**(6):861–82.

14 Bas Orth C, Schultz C, Müller CM, Frotscher M, Deller T. Loss of the cisternal organelle in the axon initial segment of cortical neurons in synaptopodin-deficient mice. *J Comp Neurol.* 2007 **504**(5):441–9.

15 Benedeczky I, Molnár E, Somogyi P. The cisternal organelle as a Ca(2+)-storing compartment associated with GABAergic synapses in the axon initial segment of hippocampal pyramidal neurones. *Exp Brain Res.* 1994 **101**(2):216–30.

16 Sanchez-Ponce D, Defelipe J, Garrido JJ, Munoz A. In vitro maturation of the cisternal organelle in the hippocampal neuron's axon initial segment. *Mol Cell Neurosci.* 2011 **48**:104–16.

17 King AN, Manning CF, Trimmer JS. A unique ion channel-clustering domain on the axon initial segment of mammalian neurons. *J Comp Neurol.* 2014 **522**(11):2594–608.

18 Kolotuev I, Schwab Y, and Labouesse M. A precise and rapid mapping protocol for correlative light and electron microscopy of small invertebrate organisms. *Biol Cell* 2010 **102**(2):121–32.

19 Kolotuev, I. Positional correlative anatomy of invertebrate model organisms increases efficiency of TEM data production. *Microsc Microanal.* 2014 **20**(5):1392–403.

20 Kukulski W, Schorb M, Welsch S, Picco A, Kaksonen M, Briggs JA. Correlated fluorescence and 3D electron microscopy with high sensitivity and spatial precision. *J Cell Biol.* 2011 **192**(1):111

21 Peddie CJ, Blight K, Wilson E, Melia C, Marrison J, Carzaniga R, et al. Correlative and integrated light and electron microscopy of in-resin GFP fluorescence, used to localize diacylglycerol in mammalian cells. *Ultramicroscopy* 2014 **143**:3–14.

22 McDonald KL. Rapid embedding methods into epoxy and LR White resins for morphological and immunological analysis of cryofixed biological specimens. *Microsc Microanal.* 2014 **20**(1):152–63.

23 Byrd DT, and Kimble J. Scratching the niche that controls Caenorhabditis elegans germline stem cells. *Semin Cell Dev Biol* 2009 **20**(9): 1107–13.

24 Cecchetelli AD, and Cram EJ. Regulating distal tip cell migration in space and time. *Mech Dev.* 2017.

25 Schalek R, Kasthuri N, Hayworth K, Berger D, Tapia J, Morgan J, et al. Development of high-throughput, high-resolution 3D reconstruction of large-volume biological tissue using automated tape collection ultramicrotomy and scanning electron microscopy. *Microsc. Microanal.* 2011 **17**, 966–67.

26 Rah JC, Bas E, Colonell J, Mishchenko Y, Karsh B, Fetter RD, et al. Thalamocortical input onto layer 5 pyramidal neurons measured using quantitative large-scale array tomography. *Front Neural Circuits* 2013 **7**:177.

27 Smith SJ, Gliko O, Serafin R, Seshamani S, MNaugle M, Parker K, et al. Fast, automated array imaging for synaptomics and connectomics. SFN abstract 2016 559.11.

28 Markert SM, Bauer V, Muenz TS, Jones NG, Helmprobst F, Britz S, et al. 3D subcellular localization with superresolution array tomography on ultrathin sections of various species. *Methods Cell Biol.* 2017 **140**:21–47.

29 van Lookeren Campagne M, Oestreicher AB, van der Krift TP, Gispen WH, Verkleij AJ Freeze-substitution and lowicryl HM20 embedding of fixed rat brain: suitability for immunogold ultrastructural localization of neural antigens. *J Histochem Cytochem* 1991 **39**:1267–79.

30 Newman GR, and Hobot JA. *Resin microscopy and on-section immunocytochemistry.* Berlin: Springer; 1993.

31 McDonald KL A review of high-pressure freezing preparation techniques for correlative light and electron microscopy of the same cells and tissues. *J Microsc* 2009 **235**:273–81.

32 Wacker I, Spomer W, Hofmann A, Thaler M, Hillmer S, Gengenbach U, et al. Hierarchical imaging: a new concept for targeted imaging of large volumes from cells to tissues. *BMC Cell Biol.* 2016 **17**(1):38.

33 Templier T, and Hahnloser RH. Automated dense collection of ultrathin sections directly onto silicon wafers. Neurosci Meet Planner San Diego, CA. Soc Neurosci Online http://www.abstractsonline.com/pp8/index.html. 2016;4071.

34 Perkovic M, Kunz M, Endesfelder U, Bunse S, Wigge C, Yu Z, et al. Correlative light- and electron microscopy with chemical tags. *J Struct Biol.* 2014 **186**(2): 205–13.

35 Paez-Segala MG, Sun MG, Shtengel G, Viswanathan S, Baird MA, Macklin JJ, et al. Fixation-resistant photoactivatable fluorescent proteins for CLEM. *Nat Methods.* 2015 **12**(3):215–18.

36 Viswanathan S, Williams ME, Bloss EB, Stasevich TJ, Speer CM, Nern A, et al. High-performance probes for light and electron microscopy. *Nat Methods* 2015 **12**(6):568–76.

37 Watanabe S, Lehmann M, Hujber E, Fetter RD, Richards J, Söhl-Kielczynski B, et al. Nanometer-resolution fluorescence electron microscopy (nano-EM) in cultured cells. *Methods Mol Biol.* 2014 **1117**:503–26.

38 Knott GW. Imaging green fluorescent protein-labeled neurons using light and electron microscopy. *Cold Spring Harb Protoc.* 2013(6):542–50.

39 Valenzuela RA, Micheva KD, Kiraly M, Li D, Madison DV. Array tomography of physiologically characterized CNS synapses. *J Neurosci Methods* 2016 **268**:43–52.

40 Jahn MT, Markert SM, Ryu T, Ravasi T, Stigloher C, Hentschel U, et al. Shedding light on cell compartmentation in the candidate phylum Poribacteria by high resolution visualization and transcriptional profiling. *Sci Rep.* 2016 **6**:35860.

41 McDonald KL. A review of high-pressure freezing preparation techniques for correlative light and electron microscopy of the same cells and tissues. *J Microsc.* 2009 **235**(3):273–81.

42 Bishop D, Nikić I, Brinkoctter M, Knecht S, Potz S, Kerschensteiner M, et al. Near-infrared branding efficiently correlates light and electron microscopy. *Nat Methods.* 2011 **8**(7):568–70.

43 Fu M, Yu X, Lu J, Zuo Y. Repetitive motor learning induces coordinated formation of clustered dendritic spines in vivo. *Nature* 2012 **483**(7387):92–5.

44 Chen SX, Kim AN, Peters AJ, Komiyama T. Subtype-specific plasticity of inhibitory circuits in motor cortex during motor learning. *Nat Neurosci.* 2015 **18**(8):1109–15.

45 Schmidt H, Gour A, Straehle J, Boergens KM, Brecht M, Helmstaedter M. Axonal synapse sorting in medial entorhinal cortex. *Nature* 2017 **549**(7673):469–75.

46 Quiroga RQ, Reddy L, Kreiman G, Koch C, Fried I. Invariant visual representation by single neurons in the human brain. *Nature* 2005 **435**(7045):1102–107.

47 Jacobs J, Weidemann CT, Miller JF, Solway A, Burke JF, Wei XX, et al. Direct recordings of grid-like neuronal activity in human spatial navigation. *Nat Neurosci.* 2013 **16**(9):1188–90.

48 Burel A, Lavault MT, Chevalier C, Gnaegi H, Prigent S, Mucciolo A, et al. A targeted 3D EM and correlative microscopy method using SEM array tomography. *Development* 2018 May 25. pii: dev.160879. doi: 10.1242/dev.160879. [Epub ahead of print] PMID: 29802150.

6

Correlative Microscopy Using Scanning Probe Microscopes

Georg Fantner[1] and Frank Lafont[2]

[1] *Laboratory for Bio- and Nano-instrumentation, School of Engineering, Interfaculty Institute of Bioengineering, Lausanne, Switzerland*
[2] *Cellular Microbiology and Physics of Infection Group, Center for Infection and Immunity of Lille, CNRS UMR8204 – Inserm U1019 – Lille Regional University Hospital Center – Institut Pasteur de Lille – Univ. Lille, France*

6.1 Introduction

Trying to observe life at the smallest scale has always motivated humans. Starting from antiquity with the development of lenses, via the first microscopes in the eighteenth century, the race to apprehend the smallest parts of matter has accelerated in the modern era. One way drew from the first X-ray spectra, in order to model what atomic-scale objects look like. Even if, using field emission electron microscopy Erwin Müller could *see* atoms from the phtalocyanine molecule, observation at the nanometrical level has remained technically challenging.

However, this barrier was broken in 1982 with Gerd Binnig and Heinrich Rohrer, who invented the scanning tunneling microscope (STM) [1], for which they received the Nobel Prize in Physics (together with Ernst Ruska for the electron microscope) in 1986. While the STM provided high resolution on conductive samples, it was not suitable for insulators and hence most biological specimens. Binnig, Quate and Gerber introduced, in 1986, the Atomic Force Microscope [2], which no longer had this limitation. In biology, the canonical building units are molecules, and soon AFM was used to image single biomolecules such as proteins [3,4] and DNA [5]. AFM can also be used to image living cells [6] and tissues [7]. In 2009, single atomic bonds inside complex molecules (submolecular level) were resolved using AFM [8]. Since then, scanning probe microscopy methods have achieved nanoscopic views with unprecedented resolution, spanning scales from the atomic all the way to the tissue level. This high resolution, however, is limited to surface characterization. Another requirement is that the sample is supported by a surface. Sample preparation is therefore an essential component for getting good AFM images, especially on biological samples such as cells [6,9].

Correlative Imaging: Focusing on the Future, First Edition. Edited by Paul Verkade and Lucy Collinson.
© 2020 John Wiley & Sons Ltd. Published 2020 by John Wiley & Sons Ltd.

The scanning probe microscopy approach has been extended from STM to a large number of imaging modalities, such as AFM, as we have seen above, but also to Kelvin probe force microscopy (KFM), scanning near-field optical microscopy (SNOM), magnetic force microscopy (MFM), scanning tunnelling potentiometry (STP), and spin polarized scanning tunnelling microscopy (SPSTM). These techniques offer the possibility to observe atoms at unprecedented scale and even to manipulate them. Indeed, in 1991 Don Eigler and his colleagues at IBM in San Jose were able to drag xenon atoms on nickel using an STM tip in order to write the company's name, establishing nanotechnology. Importantly, this technology launched the field of nanosciences.

6.2 Principles of AFM

In biology, AFM is the most frequently used near-field SPM method. Its ability to image biological samples in physiological solutions makes AFM an ideal method for the correlative approaches currently being developed in life science. In its simplest mode, AFM is based on measuring the deflection of an AFM cantilever as the cantilever is scanned over the surface. The deflection of the cantilever is kept constant via a feedback loop that adjusts the relative tip-sample height by adjusting the voltage applied to the piezoelectric ceramics that control the position of the tip or sample at high resolution. The output of the feedback controller is then a direct measure of the sample topography (Figure 6.1). In most AFMs, the detection of the cantilever is measured using the optical lever deflection method (OLD). In OLD, a laser is focused on the back of the cantilever. Depending on the angular deflection of the cantilever, the laser beam is reflected onto a different position of a position sensitive detector (PSD), often a four-quadrant photodiode (Figure 6.1a). Other means of detecting the cantilever deflection are piezoresistive self-sensing cantilevers [10,11], interferometric readout [12], or tuning fork detection [13].

The deflection of the cantilever is governed by the tip-sample forces. The effect of these tip-sample forces can be characterized by measuring a force-distance curve (Figure 6.1b). In a force-distance curve, the cantilever is first moved toward the sample until it makes contact, then the tip is pressed in further until it reaches a certain set point, after which the tip is retracted again. The typical movement of the tip during a force-distance curve shows, first, during the approach curve, a displacement toward the surface until the "jump-to-contact." At this point, van der Waals and other attractive interactions overcome the spring constant of the cantilever. Then, indentation into the sample proceeds. Once the force indentation set point is attained, the second part proceeds with the retraction curve showing the "jump-off-contact" when the tip detaches from the sample. The displacement of the cantilever will continue according to the defined ramp size limit until the next approach starts.

One of the simplest methods to obtain an AFM image is *contact mode*, where the cantilever deflection is held at a constant value by the feedback controller. This mode is easy to operate and can yield exquisitely high resolution [14]. However, the lateral forces during the scan with the tip in contact make it difficult to image soft and fragile samples. Another mode often selected uses only intermittent contact of the tip with the samples and is therefore less damaging to the tip and sample. This "intermittent contact" mode is also referred to as "TappingModeTM" mode. In tapping mode, the

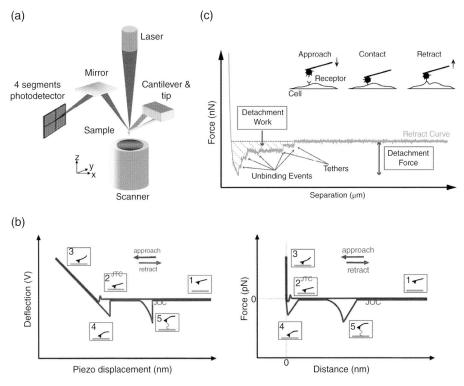

Figure 6.1 AFM principle. a: laser deflection reflected on the cantilever is detected on a four-segment photodetector. Deflection in function of the Piezo displacement (b) and force in function of the distance (c) curves depicting the cantilever movement during the approach (red curve, 1: noncontact, 2: jump-to-contact (JTC), 3: indentation) and the retraction (green curve, 4: deflection due to adhesion, 5: unbinding event and jump-off-contact). C: Retract curve with unbinding events, tethers, detachment work, and force indicated. The inset depicts the movement of the cantilever on the cell.

cantilever is oscillated close to its resonance frequency. The surface topography is detected through a change in oscillation amplitude as a function of the sample topography. A major advantage of this mode is that the lateral forces are minimized, thereby preserving the tip and the sample. Another advantage is that by monitoring the phase of the cantilever oscillation, some information can be extracted about the damping properties of the material. This concept can be further extended to multifrequency modes, where the cantilever resonance is excited and detected at multiple eigenmodes of the cantilever. By monitoring the resonance frequency, amplitude, and phase at two eigenmodes, the sample mechanical properties can be determined [15].

Recently, a new class of AFM modes has been introduced, the off-resonance tapping modes (ORT). In these modes, the cantilever is moved up and down above the sample in a controlled way, generally far below the resonance frequency of the cantilever. During this whole cycle, the cantilever deflection is recorded and the force-distance curve is extracted using real time processing. That way, the maximum interaction force during each force curve can be determined and used as the feedback parameter. Additional information about the sample can be extracted from the force curve, such as the stiffness, dissipation, and indentation depth.

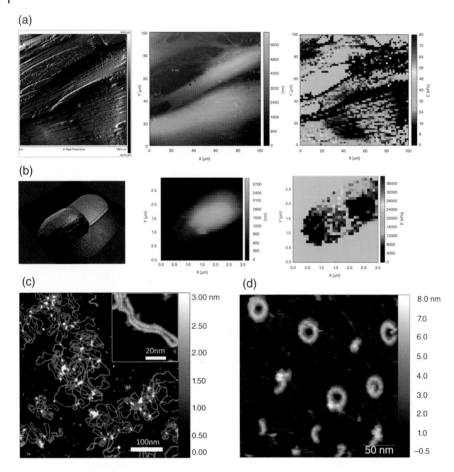

Figure 6.2 AFM imaging. a: Peak force error scan of eukaryotic HeLa cells (left), topology height image (middle) obtained in the force-volume (FV) mode and Stiffness map (right) calculated from FV. Data were obtained using a Nanowizard III AFM instrument (JPK) and PF QNM-LC cantilevers (Bruker). b: Prokaryotic *B cereus* cell deflection (3D rendering obtained using Blender, left), topology height image (middle) with the FV mode and Stiffness map (right) calculated from the FV data. Data were recorded with a MLCT cantilever (Bruker) on a Bioscope II AFM instrument (Bruker). c: DNA imaged at 20 Hz linerate using custom AFM. The inset shows the double helix of DNA. d: SAS-6 protein rings imaged in off-resonance tapping on custom AFM.

Due to its exquisite force resolution, AFM has been used in biology (mainly in liquid with soft samples), to observe soft biological objects such as cells (prokaryotic and eukaryotic, Figure 6.2a,b), viruses, single molecules and molecular complexes (e.g. DNA and the SAS-6 protein ring, Figure 6.2c,d).

AFM is not only a topography imaging technique; it gives the possibility to measure biophysics parameters such as the elastic Young's modulus during the indentation, leading to the construction of elasticity maps. It also allows us to obtain membrane viscosity information considering membrane tethers during the retraction of the cantilever [16]. Moreover, it is possible to "functionalize" the tip of the cantilever with a molecule [17] that will be used as a ligand to probe interactions with its receptor on the analyzed

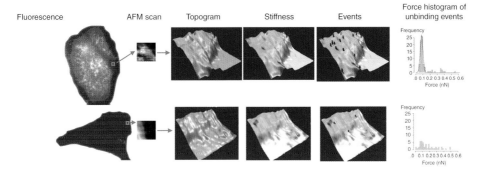

Figure 6.3 AFM parameters. Left: fluorescence grayscale of HeLa cells overexpressing a GFP-GPI construct (top) and transfectant control cell (bottom). Deflection scan (AFM scan) on a zone indicated by the orange square on the fluorescence image. Corresponding "topogram" and "stiffness" maps on the first 50 nm of indentation using the stiffness tomography approach. Cantilevers (DNP, Bruker) have been functionalized with the aerolysin toxin from *Aeromonas hydrophyla* and the functionalized tip was used to map the GFP-GPI constructs unbinding "events" indicated by the red arrows. Acquisition was obtained on a catalyst from Bruker. Interaction rupture force histograms are shown on the right.

surface. The interaction between the ligand on the tip and its receptor on the sample surface is identified on the retraction curve by the ligand/receptor rupture force (Figure 6.1b,c). This force is related to the enthalpic barrier of dissociation and depends on the loading rate (itself related to the speed of retraction and the cantilever spring constant). Hence, it is worth noting that AFM scans give access to several parameters: topography at the nm scale, elastic modulus, and providing the use of functionalized tips: distribution of the receptors and interaction forces (Figure 6.3). Adhesion forces deflecting downward the cantilever can be measured on the retraction curve, and the adhesion work can be calculated considering the area below the baseline (Figure 6.1c).

Hence, in the context of correlative microscopy related to AFM and applied in life sciences, one AFM offers correlation either with topograms or with diverse biophysics maps (e.g. adhesion, elasticity).

6.3 AFM and Optical Microscopy Correlative Approaches

The first correlation method that has emerged combining AFM and optical micros-copy consisted of placing an AFM head on an inverted microscope giving access to widefield and fluorescence microscopy. However, in the early stages, the spatiotempo-ral resolutions of the two techniques were very different. AFM scanning allowed nanometer-scale imaging at low speed, whereas optical microscopy offered video rate recording but at a resolution limit in the hundreds of nanometers. With the advent of super-resolution light microscopy, the gap between AFM and optical microscopy has narrowed, making the two methods even more compatible (see the next section). The conventional combination of wide field microscopy and AFM remains a very attractive combination, for instance, to correlate intracellular structures with extracellular topog-raphy, or to simply identify specific cells for AFM analysis. For example, the location of *Plasmodium falciparum*-infected erythrocytes was detected by fluorescently-labeled parasites, and that allowed further study of different parasite genotypes using AFM [18]. This was used to examine the structure of so-called knobs at the cell surface. These knobs result from export of parasite protein to the erythrocyte surface and play a role in the adhesion step of the infected red blood cells [19]. The knobs vary in size, depending on the parasite genotype and are in the tens of nm scale (well below the diffraction limit). For isolated molecules, such as actin, AFM allows precise analysis of clusters of actin-associated proteins such as debrin that can also be visualized at lower-resolution using TIRF [20].

6.4   Correlation with CLSM

The enhanced contrast and the available 3D resolution of confocal laser scanning microscopy (CLSM) make it a very interesting method for correlated AFM/optical microscopy. For simultaneous recording using CLSM and AFM, one has to cope with the scanning mode(s). AFM can operate either in a sample or a tip-scanning mode – the former being compatible with CLSM only if the two scanning motions are synchronized. This, however, has the big disadvantage that the temporal resolution is set by the slowest method, in this case AFM. Otherwise, sequential acquisitions allow correlation of the images, taking into account the respective dynamics. It is possible to proceed with the CLSM analysis after fixation of the samples (once the AFM is performed in living cells). In this case, CLSM is used to locate non-dynamic structure in fluorescence. For example, it has been possible to monitor the impact on the membrane of primary bone-marrow-derived mast cells of actin rearrangement and degranulation. It unveiled the presence of membrane ridges on DNP-BSA interactions (kiss-and-run model during transient granule fusion) and, showed membrane craters following granule exocytosis on cells plated on poly-Lys (kiss-and-merge model when permanent fusion occurs [21]). Using combined AFM-CLSM allows monitoring of HT1080 cell nucleus-stained deformation that can be measured using bead- or tip-functionalized cantilevers applying different forces [22]. In such a case, internal control with the chromatin decondensating agent Trichostatin A was used to impair the stiffness of nuclei. An alternative setup combining AFM and side-view optical imaging can offer the possibility to easily follow tether formation between a cell and the substrate (e.g. following the deformation of the membrane/cytoskeleton within the cell attached to the tip) or during cell-to-cell interactions [23].

6.5   Correlation with Cell Mechanics

An important biological application of AFM has been to provide a new category of marker to characterize tumor cells: stiffness, since invasive cells show a softer pattern [24] in culture plates or in biopsies [25] than healthy cells. Using an AFM combined with CLSM-based approach, it has been possible to investigate the elastic properties of cells embedded in a 3D-collagen I hydrogel [26]. The authors showed that actively invading and embedded metastatic breast adenocarcinoma cells display a Rho-associated protein kinase (ROCK)-dependent stiffening vs. cell remaining at the surface of the gel. MDA-MB-221 cell plasma membrane labeled with CellMask™ Deep-red partially embedded in the collagen matrix stained with Atto 465 NHS were analyzed using one indentation/2 μm along 40 μm at 3 μm/s speed and about 20 nN trigger force. The calculated elasticity map of apparent Young's moduli with 250 nm intervals of indentation depth was correlated to the confocal Z-axis.

AFM used to implement fluorescence strategies in order to obtain high resolution data is illustrated by the AFM/Förster energy transfer (FRET) correlation [27]. In this case, the AFM tip is functionalized with a polymer containing the acceptor dye (rhodamine), whereas the sample consists of a multilayer film of arachidonic acid incorporating the fluorescent donor dye (fluorescein). The distance dependence of the FRET only allows energy transfer from dyes in closest proximity between the sample and the tip with a resolution of approximately 400 nm. Fluorescence lifetime imaging microscopy (FLIM) gives information on the local environment affecting the fluorescence lifetime that includes solvent, quencher, energy transfer acceptor. In the Gram-negative *Shewanella oneidensis* MR-1, a correlation has been reported between the polar localization of the YFP-tagged methyl-accepting chemotaxis protein (MCP) studied by FLIM and surface protuberances analyzed using AFM [28]. Combining fluorescence correlation spectroscopy (FCS) and AFM allows us to analyze how the lipid composition impacts supported bilayers, for instance related to the line tension exerted at the line boundary of lipid domains. It also permits to study how peptides affect phase separation [29]. AFM informs on the height, the formation of dimples, and tensions whereas FCS provides information on fluorescent lipid dynamics.

6.5.1 Correlation with Super-Resolution Light Microscopy (SRLM)

Confocal resolution, depending on the wavelength and the numerical aperture of the lens, is about 200 nm (*x,y*) and 600 nm (*z*). A twofold increase can be obtained using structured illumination microscopy (SIM) that relies on a line pattern close to the diffraction limit superimposed on the specimen following several rotations/image steps. Superimposed patterns create Moiré fringes that have a lower spatial frequency than the original structure within the specimen. Using Fourier transformation, oscillatory functions are obtained from which the spatial information is extracted. SIM was introduced in 1963 [30]. Although AFM images are more defined, correlation with SIM can be used for large structures as actin tail comets (Figure 6.4a).

The gap between the achievable optical resolution and that of AFM has further decreased with the development of even higher resolution optical techniques such as stimulated emission depletion microscopy (STED) [31], and photo activated localization microscopy (PALM) [32], stochastic optical reconstruction microscopy STORM [33], and stochastic optical fluctuation imaging (SOFI) [34]. These techniques routinely achieve resolution in the few tens of nanometers.

Correlation between AFM and STED allows overlay of 20–40 nm fluorescent beads including nanomanipulation [35]. STED-AFM was applied to analyse and manipulate filaments of the cytoskeleton on fixed cells immunostained with ATTO647N- or Aberrior Star 635P-coupled antibodies against microtubules. In this approach, STED microscopy was used to locate filaments with tens of nanometers resolution and the local elasticity was measured by AFM [35–37]. Alternative to STED, the PALM/STORM methods have also been combined with AFM. Correlation has been reported between AFM-based topology and localization of Alexa Fluor 647-immunodecorated tubulin using STORM

(a)

(b)

(c)

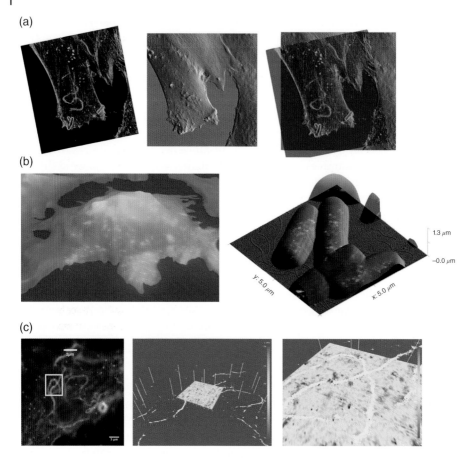

Figure 6.4 AFM correlation with SRLM. a: Left: SIM image of a PtK2 cell infected with *Shigella flexneri* (DAPI labeling, blue), actin labeled with Phalloidin-Alexa488. The image was acquired with an Elyra system (Zeiss). Middle: AFM Peak Force error image (recorded on a Catalyst AFM Instrument, Bruker). Right: merge. b: Left: rendered combined AFM PALM on HeLa cells and right: *E. coli*, recorded on custom PALM/AFM [58] (see text for details). c: HeLa cell expressing the tdEOS-actin construct and infected by *S flexneri*. Epifluorescence image (left). Right panel: perspective view of the merge images from the PALM acquisition (tdEOS-actin in yellow) and the stiffness map on the 500 nm indentation from the cell surface (indicated in yellow square in the left panel). AFM data recorded using DNP cantilevers (Bruker) with the FV mode on a catalyst AFM instrument (Bruker). The fluorescence data was obtained on an Elyra system (Zeiss). The vertical lines indicate the location of 100 nm diameter beads used for the correlation. Right panel: zoom in of the middle panel.

on fixed fibroblasts [36]. Due to the surface analysis by AFM, it was difficult to obtain elastic values for microtubules that are located deep inside the cell, or to directly compare the achievable resolutions. Correlated measurements of purified actin filaments labeled with phalloidin-ATTO488 and deposited on an APTES coated glass coverslip, allowed the direct comparison of the resolutions obtainable with AFM and STORM [38]. Time-resolved sequential AFM/STORM recorded on living CHO-K1 cell has been achieved to follow the dynamics of the cell membrane using AFM and the reorganization of paxillin-mEos2 clusters in PALM [38]. Each PALM image was created from 5,000 frames recorded at 20f/s. This allows us to monitor the formation of paxilin cluster, while the AFM permits us to follow dynamics of filopodia and lamelipodia (Figure 6.4b, left panel). AFM/PALM can also be used for prokaryotes as shown by localization of the RNA-polymerase-mEos2 in E. coli (Figure 6.4b, right panel) [38].

Key challenges in combining SRLM techniques with AFM lie in the fact that the labeling strategy for SRLM (such as antibody labeling) can adversely affect the AFM resolution, and that the laser used for the AFM cantilever detection can disturb the lifetime of the fluorophores. Special care must also be taken when correlating the images, since distortion in both the optical as well as the AFM image make a direct overlay challenging. The use of closed-loop AFM scanners and fiducial markers such as fluorescent beads are essential (Figure 6.4c). An important hurdle to overcome would also be the difficulties in using SRLM for live cell imaging. One of the strengths of AFM in biology is that it can image living cells in a time lapse manner [39]. Combining time lapse AFM with time lapse SRLM would open a broad range of important applications for this correlated technique.

6.5.2 Future Developments

Simultaneous AFM and SRLM experiments on living cells have yet to be reported. This is due to technological difficulties, such as the reflection of the excitation laser by the cantilever creating uneven illumination of the sample, the different temporal resolutions of AFM and SRLM, as well as the sometimes-incompatible imaging buffers. The first point could be addressed using the fact that the scanning in optics can be faster than in AFM. It would therefore be feasible to record optical data while the cantilever is off-contact (e.g. during the retrace). This, however, requires a level of software integration between the two instruments that currently is not available in any commercial system. The different temporal resolution is of particular concern in localization based SRLM modes. In PALM, for example, it is not uncommon to record 15 minutes of fluorophore blinking to reconstruct one PALM image. AFM generally requires a few minutes for an image, and emerging high-speed AFMs reduce this down to seconds or less. With the development of improved SRLM methods (3DPALM, PAINT, 3D-SOFI, etc.), these difficulties will hopefully be solved, eventually allowing truly simultaneous live-cell imaging by AFM and SRLM.

To make full use out of such capabilities, simultaneous correlation by synchronous registration in both modes (AFM and SRLM) will be essential in order to online analyze both physical parameter changes (from AFM) and distribution of the proteins (from SRLM) at the spatiotemporal resolution.

Another opportunity would be to establish 3D-correlation; however, the volume analyzed by the two methods is not identical. While some SRLM methods (like 3D STED) allow a complete analysis along the z-axis, others are restricted to the depth of the evanescent wave on the interface of the glass cover slip and the sample. AFM on the other hand is restricted to the top surface of the sample, allowing at best probing within the first micron below the surface of mammalian cells via indentation. It is possible to use methods to analyse differences in stiffness within the sample (e.g., the so-called stiffness tomography) [40]. Preliminary correlation between stiffness tomography and SR microscopy has been introduced [41]. It will, however, require validatation for the different organelles found intracellularly.

Quantitative biology requires us to obtain enough robust data to establish models for predictions to be validated. Due to the heterogeneity of the cell shape, it is useful to perform correlative experiments using micropatterns (Figure 6.5a) that allow averaging

(a)

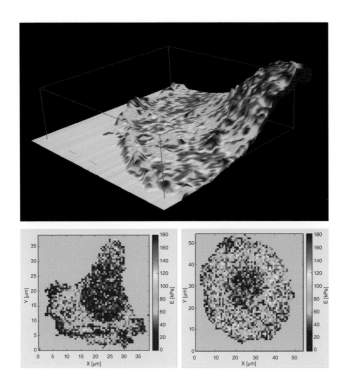

(b)

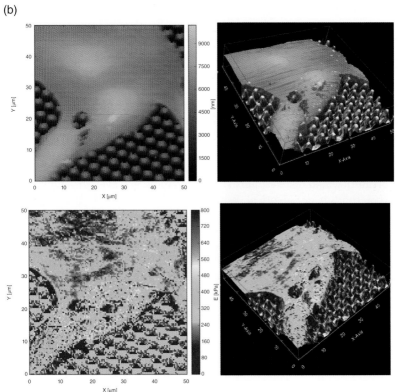

results of many cells with similar morphologies shaped by the patterned substratum. The distribution of the cytoskeleton elements can be more controlled than on conventional glass plates. Hence, correlated stiffness and molecular distribution will be more reliable, allowing challenges with drugs and gene expression modulation. Patterning can be established on glass but also on nanostructured PDMS substratum as, for instance, micropillars (Figure 6.5b). This substrate offers the advantages of analyzing force at the cell surface interface (using micropillars displacement in correlation with labeling of the cytoskeleton and/or the focal adhesion plaques molecular machinery) and at the cell cortex using AFM. In order to analyze more cells, microfluidic systems can also be envisioned that will allow regularly spaced cell deposition with the possibility of individual treatments. This can also be combined with micropatterned substrates. Peric et al. have demonstrated a multilayer microfluidic device for repetitive trapping of bacteria for correlated AFM/optical microscopy imaging [42].

Alternatively, correlative measurements can be performed on tissue sections possibly obtained as biopsies from patients. It has thus been possible to establish a correlation between the invasiveness of cancer cells and their stiffness, by comparing stiffness maps of normal, benign and invasive cancer tissues obtained from human breast cancer [25]. Malignant tissues display more heterogeneity, with stiffness varying from the core to the periphery in biopsies processed for histopathology examination. Extracellular matrix nanomechanical properties in cancer at different stages can also be analyzed using this approach. Similarly, ECM deposition can increase corneal stroma stiffness in notch1-deficient corneas (a model of chronic inflammation) that correlate with an increase in tenascin C expression, with effect at the limbus, peripheral and central corneal epithelium [43]. Hence, in the future, one has to consider that more studies will be performed directly on tissues from biopsies adding biophysical markers as elasticity among the biomarkers for diagnosis purposes. This point can also apply for isolated cells.

6.6   AFM and Correlation with Electron Microscopy

Both AFM and electron microscopy (transmission (TEM) and scanning (SEM)) achieve resolution in the (sub-)nanometer scale. The specific advantage of TEM is that it can look through samples if they are sufficiently thin. It is, therefore, often used in biology to look at intracellular structures. SEM images have the advantage that they are very intuitive with respect to interpretation. For both techniques, additional sources of information exist such as elemental analysis using energy dispersive X-ray (EDX). The samples, however, have to be prepared with several harsh treatments (in the case of TEM, dehydration and often staining). In SEM, samples also have to be dehydrated and

Figure 6.5 AFM stiffness maps on cells plated on micropatterns (a) and micropillars (b). A: Stiffness maps of HeLa cells plated on crossbow (bottom left) and disk (bottom right) micropatterns. Maps were obtained using Biolever mini-cantilevers (Olympus) and recorded with the QI-mode (JPK) on a nanowizard III (JPK) instrument. Top: 3D rendering of a part of the cell bottom right. B: Ptk2 cells plated on micropillars (deflection image top, stiffness maps bottom). Data were recorded using PF QNM-LC cantilevers (Bruker) in the QI mode (JPK) on a Nanowizard III equipped with cellhesion AFM instrument (JPK).

metal- or carbon-coated if the sample isn't intrinsically conductive. An exception is environmental SEM, where the samples are imaged at 100% humidity at temperatures around zero centigrade. However, the resolution of these systems is significantly lower than conventional high vacuum SEM.

While the lateral resolutions of SEM, TEM and AFM are roughly comparable, their Z-resolutions are vastly different. SEM has a depth resolution dependent on the interaction volume, and in the case of TEM tomography the third dimension can be back-calculated by taking TEM images from multiple angles. The Z-resolution is therefore lower than in X and Y. In AFM, however, the Z-resolution is generally an order of magnitude higher than in X and Y. With its ability to measure distances accurately in three dimensions, measure mechanical and electrical properties, and use for nanomanipulation, the AFM therefore provides much value when correlated to electron microscopy.

Correlation *stricto sensu* can be achieved, for instance, by having an AFM setup within the chamber of a SEM providing information on the 3D topography, mechanical, and electrical properties of the sample analyzed under the electron beam. Most biological applications are based on sequential correlative analysis with AFM performed on living or isolated and unprocessed samples. As an example of the latter case, analysis of human saliva exosomes was reported using AFM and field emission scanning electron microscopy (FESEM). AFM permits monitoring of the nanomechanical properties of exosomes. Measuring interactions with membrane markers using functionalized tips allows us to define the endosomal instead of plasma membrane origin, while FESEM ascertains the globular shape of the exosomes [44].

Using multiple correlative microscopy steps can give rise to complex experimental procedures. For instance, to analyse the influence of osteoclast adhesion to cortical bone surface, anchor points on the bone surface are examined along the following sequence of correlated acquisitions: SEM analysis in air of the bone sample before cell plating, live fluorescence of the cells before fixation and SEM analysis in air of the cell/bone sample, then, after removal of the cell (using NaOCl short treatment followed by drying under nitrogen stream) post cell removal bone is analyzed by environmental SEM and AFM [45]. Through all these steps, correlation has to be maintained. This can be done using always the same orientation of the sample holder throughout the sequence of analysis and specific points on the bone. However, this requires that all instruments can accommodate the same type of sample holder. Systems like this are becoming commercially available, such as from Hitachi High Technologies Corporation. An alternative approach to obtain confident correlation between images produced by different methods is the use of fiducial markers (such as beads), especially when the resolution of the techniques is in the same order, for instance, AFM, SRLM, and EM as previously described in the CLAFEM method [41] illustrated in Figure 6.6 examining mitochondria.

6.6.1   Correlation Involving AFM, EM, and Chemical Surface Characterization

We have mentioned that another layer of analysis can be added combining microscopy methods to chemical analysis techniques. Confocal Raman microspectroscopy can be combined with AFM to analyze, on the one hand, difference in membrane chemical composition of several cancer cells and, on the other hand, their nanomechanical

(a)

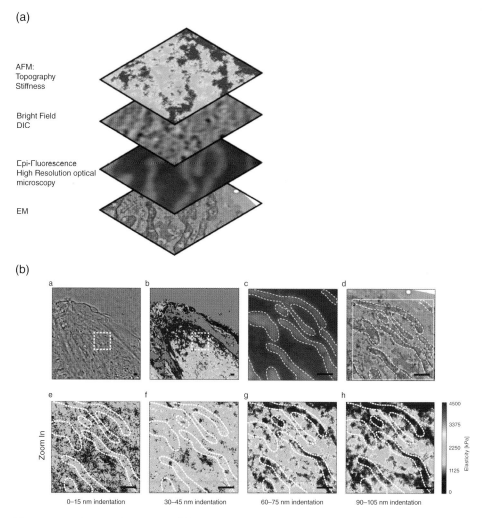

AFM:
Topography
Stiffness

Bright Field
DIC

Epi-Fluorescence
High Resolution optical
microscopy

EM

(b)

Zoom In

0–15 nm indentation   30–45 nm indentation   60–75 nm indentation   90–105 nm indentation

Elasticity [kPa]

4500

3375

2250

1125

0

Figure 6.6 a: CLAFEM description of the possible correlated modes of acquisition on HeLa cell labeled with mitotracker: AFM (incl. topogram, stiffness mapping (shown in pseudo-color)), bright field (incl. DIC), fluorescence (incl. epifluorescence (shown), SROM), electron microscopy (incl TEM (shown), SEM). b: Example of CLAFEM: a: bright field of a part of a HeLa cell, b: corresponding elasticity map c: mitotracker labeling with mitochondria underlined with white dotted lines, d: TEM image with the panel c indicated within the white square (mitochondria shown within white dotted lines as in panel c), e–f: elasticity maps corresponding to the panel c at different indentations (indicated below the maps). Elasticity colour scale is indicated varying from blue (soft) to red (stiff).

features [46]. Combination can include electron microscopy. For example, to analyze plant cell wall architecture, inhomogeneity of layering structures of sclerenchymatic fibers can be analyzed by TEM, whereas AFM imaging informs on the order of the cellulose microfibrils deposited in the primary wall of the protoxylem vessel, and, finally Raman images tells about the microfibril orientation [47]. Another approach involves a short-distance (10 nm) analysis depth technique enabling surface chemistry characterization,

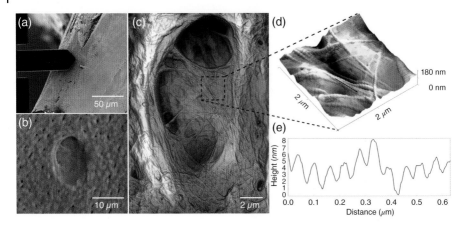

Figure 6.7 *In situ* correlated SEM/AFM imaging of a collagen lined lacunae in bovine trabecular bone. a: AFM integrated inside the SEM allows for direct, high depth of focus viewing of both tip and sample. b: using the large field of view and fast image acquisition of the SEM it is possible to find rare and small features like lacunae and the AFM cantilever can be positioned there even on very irregular surfaces. c: AFM overview image of the lacunae showing the collagen fibers. d: High-resolution image of the collagen fibers showing the characteristic 67 nm periodic banding pattern. e: *X-Z* cross section of the banding pattern revealing a corrugation height of 3 nm. Images were taken using the AFSEMTM instrument (GETec GesmbH) integrated in a Quanta field emission SEM (FEI). Image courtesy Dr. Marcel Winhold.

like time-of-flight secondary ion mass spectrometry (ToF-SIMS) for correlative study with SEM and AFM [48]. In this case, various organic species of nucleolus-like body units of mammalian oocytes freeze-dried and embedded in the epoxy resin without fixation were analyzed using ToF-SIMS. The resolution is limited by the ion beam (size of pixel about 0.2 μm), which allows limited but significant correlation with the topology of the structures analyzed with AFM and the ultrastructural data of large field obtained with SEM.

The area in which correlated AFM/EM microscopy is the most integrated is the combination of AFM with SEM. Here, dedicated solutions exist where both imaging modalities are integrated within one instrument. The advantage is that it becomes easy to apply different methods *in situ* at the exact same spot of the sample. This is particularly advantageous when the sample is extremely small or difficult to reach with the AFM cantilever, for example, the case of a bovine vertebra as shown in Figure 6.7, where the precision of the SEM was required to guide the AFM cantilever onto the single trabecula. Taking this even further, Kreith et al. have reported the use of combined AFM, nanoindentation and SEM imaging where they measured the appearance and height of slip step dislocations on a single crystal copper micro-pillar. The AFM inside the SEM can however also be used for manipulation rather than imaging. Fukushima et al have used an AFM inside an SEM to measure the forces required to break a 3D structure [49]. Iwata et al. have used an AFM inside an SEM to perform precision small scale nanomanipulation using a haptic interface [50]. The combination of AFM with SEM becomes even more interesting when paired with the myriad of other modes and techniques that are available in SEM (EDX, FIB, four-point probing, mechanical stretching stages, etc.) or AFM (conductive AFM, nanomechanical mapping, magnetic force microscopy, etc.).

6.6.2 Future Developments

Tomography methods are spreading in the SEM community with focused ion beam SEM (FIB SEM) and serial block-face (SBF SEM) that can be used for correlative microscopy with fluorescence, including SRLM. One caveat is that the reconstruction still remains full of uncertainties and is highly time-consuming. Several approaches are developed to allow automated detection of membranes within cellular scans, and among those approaches, AFM has been tried as an accurate topography analysis tool to provide profiles that can be used for reconstruction purposes. The method is not yet mature but might be an alternative to the image analysis possibilities currently explored or may complement them. Also, these volume SEM approaches can be correlated with the 3D-SR optical microscopy (3D-PALM, 3D-STED,...) and the stiffness tomography obtained using AFM. This would then allow us to test whether diet, drugs, and metabolic disorders affect the mechanics of intracellular organelles – for instance, depending on the expression/ distribution of molecular markers.

Most correlative analysis methods involving AFM include techniques where the different imaging modalities were done in different instruments (except for the case of AFM/optical microscopes). Therefore, almost all correlated experiments are performed sequentially. Only a few examples are available for simultaneous recording. The two main reasons are the integration and the synchronization. Integration of optical and electron microscopy, AFM and optical *or* electron microscopy has been reported. To date, however, all three techniques have not been yet integrated in a single setup. Whether this will happen in the coming years remains highly conjectural. Indeed, the most probable achievement will be devoted to integrate high-speed AFM with the other microscopy techniques.

6.7 Future Developments Involving Correlation Microscopy Using HS-AFM

In 2001, Ando and co-workers introduced the HS-AFM that allows us to analyze the structural dynamics of molecules. HS-AFM was released with the visualization at unpreceded resolution and dynamics of the walking-like movement of *machine-at-work* as molecular motors on cytoskeleton filaments. Also, emerged the possibility to record *cellular* processes at seconds to sub-100 ms resolution. This is achieved on unprocessed samples hence not disturbing their functions [51]. Since then, several important breakthroughs have been obtained mainly analyzing recording molecular processes. For instance, it has been possible to visualize the assembly of the snip7 component of the ESCRTIII establishing spirals on supported bilayers that could explain in part the formation together with the presence of the cargo of budding endosomes [52,53]. High-speed imaging of living cells has also been reported and even integration of HS-AFM with basic optical microscopy [54] as well as advanced optical microscopy [38] has been achieved. One can clearly envision that the integration of the set-ups would allow synchronous force and fluorescence analysis before very rapidly freezing the sample for further processing in electron microscopy. Also, improvements in off-resonant tapping modes should render it feasible to measure elastic properties of the samples at high speed, while performing fluorescence measurements. For recognition imaging this might be more difficult, since a minimal interaction time is required between the ligand and the receptor. The use of HS-AFM with functionalized tips to obtain interaction force maps may be more challenging.

Another aspect of the development of HS-AFM is that it drastically reduces the image acquisition time. This is not only important when trying to resolve changes of the sample but also has a big impact on the practical use of correlated microscopy techniques. Especially in truly combined systems, such as AFM/SEM combinations, the long acquisition time of conventional AFM is in stark contrast to the short time it takes to obtain an SEM image. Using HS-AFM in such situations significantly improves the workflow as well as reduces imaging costs, given that high-end SEM systems often exist in user facilities where the systems are in high demand.

The main drawback of HS-AFM is the limited scan range. HS-AFM scanners with larger area do exist [55,56], however, there is a trade-off between scan size and scan speed. Video-rate AFM in tapping mode can thus far only be achieved in solutions, due to the higher Q-factor of cantilevers when they are operated in air or vacuum. Using unconventional materials for the AFM cantilevers has shown potential in this area, enabling HS-AFM imaging in tapping mode also in air [57] and potentially vacuum. It is to be expected that this technology will also be used in correlated AFM systems, thereby bridging the gap in imaging speeds between, for example, AFM and SEM.

6.8 Concluding Remarks

The correlation of AFM with optical microscopy was recognized early on as a natural match, due to the complementary information that these techniques can provide. Soon after, the combination with electron microscopes was demonstrated to have great potential. Other, more specialized correlated measurement systems have also been reported. What most of these systems have in common, however, is that the focus has been on enhancing the AFM image with other information. Hence, the focus has traditionally been on the AFM instrument, and AFM users. We can only speculate on why this is the case. Maybe it is because operating an AFM is considered relatively complex and still lacks the level of user-friendliness other microscopy techniques such as optical or electron microscopy provide. With ongoing development in the area of AFM automation and ease of use, this is likely to change. At that point, the focus can move away from the AFM, and rather move towards fully integrated workflows involving all imaging modalities. This will allow a much broader use of the richness of information accessible through correlated microscopy.

Acknowledgments

We apologise for not citing numerous authors and their valuable and relevant contributions to the field due to space limitation. We would like to thank members of the CMPI group, S. Janel, Drs. M. Popoff, V. Dupres for performing correlated imaging (Figures 6.3, 6.4, 6.6), elasticity mapping (Figures 6.2, 6.5) and help for Figure 6.1, and Dr. Marcel Winhold, GETec GesmbH for performing the in situ correlated AFM/SEM measurements. This work has been supported by grants from the ANR (10-EQPX-04-01) and the EU-FEDER (12,001,407) to FL.

References

1 Binnig G, and Rohrer H. Scanning tunneling microscopy. *Surf. Sci.* 1983 **126**:236–44.

2 Binnig G, Quate CF, and Rohrer C. Atomic force microscope. *Phys. Rev. Lett.* 1986 **56**:930–33.

3 Edstrom RD, Yang XR, Lee G, and Evans DF. 1990. Viewing molecules with scanning tunneling microscopy and atomic force microscopy. *FASEB J.* 1990 **4**:3144–51.

4 Weisenhorn AL, Drake B, Prater CB, Gould SA, Hansma PK, Ohnesorge F, Egger M, Heyn SP, Gaub HE. Immobilized proteins in buffer imaged at molecular resolution by atomic force microscopy. *Biophysical journal.* 1990 Nov 1;**58**(5):1251–8.

5 Lindsay SM, Nagahara LA, Thundat T, Knipping U, Rill RL, Drake B, Prater CB, Weisenhorn AL, Gould SA, Hansma PK. STM and AFM images of nucleosome DNA under water. *Journal of Biomolecular Structure and Dynamics.* 1989 Oct 1;**7**(2):279–87.

6 Henderson E, Haydon PG, and Sakaguchi DS. Actin filament dynamics in living glial cells imaged by atomic force microscopy. *Science* 1992 *(80-.)***257**:1944–46.

7 Tao NJ, Lindsay SM, and Lees S. Measuring the microelastic properties of biological material. *Biophys. J.* 1992 **63**:1165–69.

8 Gross L, Mohn F, Moll N, Liljeroth P, and Meyer G. The Chemical Structure of a Molecule Resolved by Atomic Force Microscopy. *Science* 2009 *(80-.).* **325**:1110–14.

9 Kasas S, Gotzos V, and Celio MR. Observation of living cells using the atomic force microscope. *Biophys. J.* 1993 **64**:539–44.

10 Tortonese M, Barrett RC, and Quate CF. Atomic resolution with an atomic force microscope using piezoresistive detection. *Appl. Phys. Lett.* 1993 **62**:834–36.

11 Dukic M, Adams JD, and Fantner GE. Piezoresistive AFM cantilevers surpassing standard optical beam deflection in low noise topography imaging. *Sci. Rep.* 2015 **5**:16393.

12 Hoogenboom BW, Frederix P, Yang JL, and Martin, S. A Fabry-Perot interferometer for micrometer-sized cantilevers. *Appl. Phys. Lett.* 2005 **86**:1–3.

13 Giessibl FJ. Atomic resolution on Si(111)-(7×7) by noncontact atomic force microscopy with a force sensor based on a quartz tuning fork. *Appl. Phys. Lett.* 2000 **76**:1470.

14 Müller DJ, Schoenenberger C-A, Schabert F, and Engel A. Structural Changes in Native Membrane Proteins Monitored at Subnanometer Resolution with the Atomic Force Microscope: A Review. *J. Struct. Biol.* 1997 **119**:149–57.

15 Herruzo ET, Perrino AP, and Garcia R. Fast nanomechanical spectroscopy of soft matter. *Nat. Commun.* 2014 **5**:3126.

16 Müller D, Helenius J, Alsteens D, Dufrêne Y, Müller DJ, Helenius J, et al. Force probing surfaces of living cells to molecular resolution. *Nat. Chem. Biol.* 2009 **5**(6), 383–90.

17 Ebner A, Wildling L, Zhu R, Rankl C, Haselgrübler T, Hinterdorfer P, Gruber HJ. Functionalization of probe tips and supports for single-molecule recognition force microscopy. *InSTM and AFM Studies on (Bio) Molecular Systems: Unravelling the Nanoworld* 2008 (pp. 29–76). Springer, Berlin, Heidelberg.

18 Nacer A, Roux E, Pomel S, Scheidig-Benatar C, Sakamoto H, Lafont F, Scherf A, Mattei D. Clag9 is not essential for PfEMP1 surface expression in non-cytoadherent Plasmodium falciparum parasites with a chromosome 9 deletion. *PLOS ONE.* 2011 Dec 19;**6**(12):e29039.

19 Sharma, YD. Knob proteins in falciparum malaria. *Indian Journal of Medical Research* 1997 **106**:53–62.

20 Sharma S, Grintsevich EE, Hsueh C, Reisler E, and Gimzewski JK. Molecular cooperativity of drebrin1-300 binding and structural remodeling of F-actin. *Biophys. J.* 2012 **103**:275–83.

21 Deng Z, Zink T, Chen HY, Walters D, Liu FT, Liu GY. Impact of actin rearrangement and degranulation on the membrane structure of primary mast cells: a combined atomic force and laser scanning confocal microscopy investigation. *Biophysical journal.* 2009 Feb 18;**96**(4):1629–39.

22 Krause M, te Riet J, and Wolf K. Probing the compressibility of tumor cell nuclei by combined atomic force–confocal microscopy. *Phys. Biol.* 2013 **10**:65002.

23 Chaudhuri O, Parekh SH, Lam WA, and Fletcher DA. Combined atomic force microscopy and side-view optical imaging for mechanical studies of cells. *Nat. Methods* 2009 **6**:383–7.

24 Lekka M, Laidler P, Gil D, Lekki J, Stachura Z, Hrynkiewicz AZ. Elasticity of normal and cancerous human bladder cells studied by scanning force microscopy. *European Biophysics Journal.* 1999 May 1;**28**(4):312–6.

25 Plodinec M, Loparic M, Monnier CA, Obermann EC, Zanetti-Dallenbach R, Oertle P. The nanomechanical signature of breast cancer. *Nature nanotechnology.* 2012 Nov;**7**(11):757.

26 Staunton JR, Doss BL, Lindsay S, and Ros R. Correlating confocal microscopy and atomic force indentation reveals metastatic cancer cells stiffen during invasion into collagen I matrices. *Sci. Rep.* 2016 **6**:19686.

27 Vickery SA, and Dunn RC. Combining AFM and FRET for high resolution fluorescence microscopy. *J. Microsc.* 2001 **202**:408–12.

28 Micic M, Hu D, Suh YD, Newton G, Romine M, and Lu HP. Correlated atomic force microscopy and fluorescence lifetime imaging of live bacterial cells. *Colloids and Surfaces B: Biointerfaces.* 2004 Apr 15;**34**(4):205–12.

29 Chiantia, S, Kahya, N, Ries, J and Schwille, P. Effects of Ceramide on Liquid-Ordered Domains Investigated by Simultaneous AFM and FCS. *Biophys. J.* 2006 **90**:4500–508.

30 Lukosz W, and Marchand M. Optischen Abbildung Unter Überschreitung der Beugungsbedingten Auflösungsgrenze. *Opt. Acta Int. J. Opt.* 1963 **10**:241–55.

31 Hell SW, and Wichmann J. Breaking the diffraction resolution limit by stimulated emission: stimulated emission depletion microscopy. *Opt. Lett.* 1994 **19**:780–82.

32 Betzig E, Patterson GH, Sougrat R, Lindwasser OW, Olenych S, Bonifacino JS, et al. Imaging intracellular fluorescent proteins at nanometer resolution. *Science.* 2006 Sep 15;**313**(5793):1642–5.

33 Rust MJ, Bates M, and Zhuang XW. Sub-diffraction-limit imaging by stochastic optical reconstruction microscopy (STORM). *Nat Methods* 2006 **3**:793–95.

34 Dertinger T, Colyer R, Iyer G, Weiss S, Enderlein J. Fast, background-free, 3D super-resolution optical fluctuation imaging (SOFI). *Proceedings of the National Academy of Sciences.* 2009 Dec 29;**106**(52):22287–92.

35 Harke B, Chacko JV, Haschke H, Canale C, and Diaspro A. A novel nanoscopic tool by combining AFM with STED microscopy. *Opt. Nanoscopy* 2012 **1**, 3.

36 Chacko JV, Zanacchi FC, and Diaspro A. Probing cellular structures by coupling optical super resolution and atomic force microscopy techniques for a correlative approach. *Cytoskeleton (Hoboken).* 1–19 (2013). doi:10.1002/cm.

37 Chacko JV, Harke B, Canale C, and Diaspro A. Cellular level nanomanipulation using atomic force microscope aided with superresolution imaging. *J. Biomed. Opt.* 2014 **19**: 105003.

38 Odermatt PD, Shivanandan A, Deschout H, Jankele R, Nievergelt AP, Feletti L, et al. High-resolution correlative microscopy: bridging the gap between single molecule localization microscopy and atomic force microscopy. *Nano letters.* 2015 Jul 6;**15**(8):4896–904.

39 Eskandarian HA, Odermatt PD, Ven JX, Hannebelle MT, Nievergelt AP, Dhar N, et al. Division site selection linked to inherited cell surface wave troughs in mycobacteria. *Nature microbiology.* 2017 Sep;**2**(9):17094.

40 Roduit C, van der Goot FG, De Los Rios P, Yersin A, Steiner P, Dietler G, et al. Elastic membrane heterogeneity of living cells revealed by stiff nanoscale membrane domains. *Biophysical journal.* 2008 Feb 15;**94**(4):1521–32.

41 Janel S, Werkmeister E, Bongiovanni A, Lafont F, and Barois N. CLAFEM: Correlative light atomic force electron microscopy. *Methods Cell Biol.* 2017 **140**: 165–85.

42 Peric O, Hannebelle M, Adams JD, and Fantner GE. Microfluidic bacterial traps for simultaneous fluorescence and atomic force microscopy. *Sci. Rep.* 2017.

43 Nowell CS, Odermatt PD, Azzolin L, Hohnel S, Wagner EF, Fantner GE, et al. Chronic inflammation imposes aberrant cell fate in regenerating epithelia through mechanotransduction. *Nature cell biology.* 2016 Feb;**18**(2):168.

44 Sharma S, Rasool HI, Palanisamy V, Mathisen C, Schmidt M, Wong DT, et al. Structural-mechanical characterization of nanoparticle exosomes in human saliva, using correlative AFM, FESEM, and force spectroscopy. *ACS nano.* 2010 Mar 10;**4**(4): 1921–6.

45 Shemesh, M., Addadi, S., Milstein, Y., Geiger, B. and Addadi, L. Study of Osteoclast Adhesion to Cortical Bone Surfaces: A Correlative Microscopy Approach for Concomitant Imaging of Cellular Dynamics and Surface Modifications. *ACS Applied Materials and Interfaces* 2016 **8**:14932–43.

46 McEwen GD, Wu Y, Tang M, Qi X, Xiao Z, Baker SM, et al. Subcellular spectroscopic markers, topography and nanomechanics of human lung cancer and breast cancer cells examined by combined confocal Raman microspectroscopy and atomic force microscopy. *Analyst.* 2013;**138**(3):787–97.

47 Ma J, Lv X, Yang S, Tian G, and Liu X. Structural Insight into Cell Wall Architecture of Micanthus sinensis cv. using Correlative Microscopy Approaches. *Microsc. Microanal.* 2015 1–10. doi:10.1017/S1431927615014932

48 Gulin A, Nadtochenko V, Astafiev A, Pogorelova V, Rtimi S, Pogorelov A. Correlating microscopy techniques and ToF-SIMS analysis of fully grown mammalian oocytes. *Analyst.* 2016;**141**(13):4121–9.

49 Yoshikawa S, Murata R, Shida S, Uwai K, Suzuki T, Katsumata S, et al. Evaluation of correlation between dissolution rates of loxoprofen tablets and their surface morphology observed by scanning electron microscope and atomic force microscope. *Chemical and pharmaceutical bulletin.* 2010 Jan 1;**58**(1):34–7.

50 Iwata F, Mizuguchi Y, Ko H, and Ushiki T. Nanomanipulation of biological samples using a compact atomic force microscope under scanning electron microscope observation. *J. Electron Microsc. (Tokyo).* 2011 **60**:359–66.

51 Ando T. et al. A high-speed atomic force microscope for studying biological macromolecules. *Proc. Natl. Acad. Sci. U. S. A.* 2001 **98**:12468–72.

52 Chiaruttini N, Redondo-Morata L, Colom A, Humbert F, Lenz M, Scheuring S, et al. Relaxation of loaded ESCRT-III spiral springs drives membrane deformation. *Cell.* 2015 Nov 5;**163**(4):866–79.

53 Mierzwa BE, Chiaruttini N, Redondo-Morata L, von Filseck JM, König J, Larios J, et al. Dynamic subunit turnover in ESCRT-III assemblies is regulated by Vps4 to mediate membrane remodelling during cytokinesis. *Nature cell biology.* 2017 Jul;**19**(7):787.

54 Colom A, Casuso I, Rico F, and Scheuring, S. A hybrid high-speed atomic force–optical microscope for visualizing single membrane proteins on eukaryotic cells. *Nat. Commun.* 2013 **4**:1–8.

55 Nievergelt AP, Erickson BW, Hosseini N, Adams JD, and Fantner GE. Studying biological membranes with extended range high-speed atomic force microscopy. *Sci. Rep.* 2015 **5**:11987.

56 Watanabe H, Uchihashi T, Kobashi T, Shibata M, Nishiyama J, Yasuda R, et al. Wide-area scanner for high-speed atomic force microscopy. *Review of Scientific Instruments.* 2013 May 3;**84**(5):053702.

57 Adams JD, Erickson BW, Grossenbacher J, Brugger J, Nievergelt A, Fantner GE. Harnessing the damping properties of materials for high-speed atomic force microscopy. *Nature nanotechnology.* 2016 Feb;**11**(2):147.

58 Odermatt PD, Shivanandan A, Deschout H, Jankele R, Nievergelt AP, Feletti L, et al. High-resolution correlative microscopy: bridging the gap between single molecule localization microscopy and atomic force microscopy. *Nano letters.* 2015 Jul 6;**15**(8):4896–904.

7

Integrated Light and Electron Microscopy

R. I. Koning[1]*, A. Srinivasa Raja[2]*, R. I. Lane[2]*, A. J. Koster[1], and J. P. Hoogenboom[2]

[1] *Cell and Chemical Biology, Leiden University Medical Center, The Netherlands*
[2] *Imaging Physics, Delft University of Technology, The Netherlands*

7.1 Introduction

In the past decades, correlative light and electron microscopy (CLEM) methods have evolved from being mostly used by a few pioneering, specialist labs to a collection of techniques and workflows practiced by a broad group of researchers in structural biology [1]. In most cases, CLEM involves a distinct set of sequentially used specimen preparation and labeling techniques, followed by diverse types of light and electron microscopy techniques, with specific workflows for sample transfer and relocation of regions of interest (ROIs). A key advantage of sequential CLEM is the wide diversity of available microscopes: in principle, any type of microscope can be added to the workflow, provided requirements on sample preparation and handling can be met. However, workflow procedures to combine different (light and electron) microscopes can be tedious, involving extensive manual labor, transfers, and relocation of regions of interest (ROIs), which can be cumbersome and prone to errors. Microscopes that integrate a light and an electron microscope in one have been developed as early as the 1980s and a wide variety of integrated microscopes with different modalities has been reported in literature in recent years [2, 3]. Several of these have now also become commercially available.

For a specific CLEM experiment, the choice between an experimental workflow with standalone microscopes or with an integrated microscope depends on a variety of factors, including the precise goal and requirements of the experiment, amenable sample preparation protocols, and local availability of microscopes, probes, and expertise. If only a single or very few samples have to be carried through the CLEM workflow, adopting sample preparation protocols towards integrated inspection may be an effort that does not outweigh the potential benefits. However, in terms of throughput, avoiding sample contamination, achievable precision of ROI retrieval, and ease and accuracy of image correlation, integrated microscopes offer advantages. In this chapter, we will

---

* *These authors contributed equally to this work.*

focus on those areas in present-day CLEM that are faced with challenges for which these advantages of integrated microscopes may well be key for further advancement. These areas, in our opinion, are large-scale and high-throughput correlated (volume) microscopy, super-resolution localization in resin or cryo-frozen sections, fluorescence-guided FIB milling for cryo-electron tomography, and the integration of sample preparation and transfer. Ultimately this should lead to the development of specific integrated CLEM systems with complete and fully automated workflows, leading to high-throughput and high-yield systems.

7.2 Large-Scale and High-Throughput (Volume) Microscopy

7.2.1 Advantages and Challenges for Large-Scale EM

There are a variety of techniques for three-dimensional imaging by electron microscopy of biological specimens. Modern volume EM techniques can be divided into two broader methods: array tomography-based approaches in which ultrathin serial sections are cut from a block of tissue prior to EM imaging, such as serial section TEM (ssTEM) and SEM (ssSEM), and blockface-based approaches in which the tissue block is sectioned as it is imaged, such as serial blockface SEM (SBF-SEM) and focused ion beam SEM (FIB-SEM). While both of these methods have their respective advantages and disadvantages [4–6], an important distinction is that array tomography allows for re-evaluation of sections whereas the specimen is irrevocably lost in blockface approaches.

While these techniques have been successful at generating high-quality 3D reconstructions, significant challenges remain. At present, one of the most stringent constraints facing volume EM imaging with pixel size <10 nm is throughput. To image with SEM an entire mouse cortical column, for example, of size 400 μm (X) by 1000 μm (Y) by 400 μm (Z) at 4 nm/px and 30 nm section thickness (which is the resolution scale necessary to reliably detect certain subcellular structures), Briggman and Bock have estimated that it would take approximately 500 days of uninterrupted imaging [4]. For this reason, it is advantageous to locate regions of interest prior to large-scale EM imaging to minimize the imaging volume.

One approach, taken by Hildebrand et al. for whole-brain ssSEM reconstruction of a larval zebrafish, was to utilize multiple rounds of targeted EM imaging at successively higher levels of magnification [7]. Similar approaches have likewise been taken in other large-scale neural reconstruction endeavors such as Bock et al. in partial brain imaging of a mouse visual cortex [8] and Zheng et al. in full brain imaging of an adult *Drosophila melanogaster* [9]. Although these multiscale approaches have been implemented with great success, throughput remains a bottleneck primarily due to the need for intermediate analysis of the EM dataset. The selection of subregions of interest for imaging at successive magnification scales is driven by localization of the biological material of interest, which can only be done after manual or machine-learning-assisted analysis of the preceding EM dataset. While machine-learning techniques have made tremendous progress in reducing human involvement, interpretation and annotation of EM datasets remain a tedious and error-prone practice. These methods are therefore not yet appropriate for selecting subregions at higher magnification scales, meaning selection cannot be done either automatically or in real time.

7.2.2   Advantages of CLEM for Large-Scale EM

EM imaging has the additional limitation that it does not contain the protein- and molecular-specific information available from fluorescence microscopy (FM), unless the proteins are known to be specifically linked to a structural component. This information is not only crucial for understanding biological function but can also be used to guide to ROIs based on molecular expression. Thus, while EM is successful at providing ultrastructural information, it is not always useful for localizing the ROIs. Functional fluorescence microscopy has been employed to identify the biological material of interest, mostly for blockface approaches [10], but workflows to retrieve selected regions from the specimen and trim the block to the appropriate size can be both complicated and time-consuming. Additionally, they may involve further rounds of multimodal inspection, e.g. with X-ray tomography [11]. Finally, when only EM imaging is performed after the extensive sample preparation following FM it may be very hard to link between the structural information conveyed by EM and dynamic, functional data obtained with live-cell FM.

7.2.3   Prospects for Integrated Microscopy

Integrated array tomography seeks to build on the success and usefulness of modern volume EM imaging techniques by addressing these challenges. In conventional array tomography, a tissue specimen is chemically fixed, embedded in resin, and cut into a series of ultrathin sections which are collected as ribbons on a solid substrate or on flexible, sticky Kapton tape. The sections are then immunostained for imaging in a widefield fluorescence microscope, possibly in several rounds to highlight multiple molecules, and finally heavy metal stained for EM imaging [12, 13]. Integrated array tomography combines FM and EM image acquisition with high alignment accuracy through the use of an integrated microscope. Further, as this requires samples with both fluorescence labeling and EM staining present, it removes the need for intermediate sample preparations that carry the risk of distorting the sample. A visual representation of a hypothetical workflow for integrated correlative array tomography (iCAT) is shown in Figure 7.1. In principal, such a workflow begins by following a customized protocol for fixation, embedding, and staining the sample such that fluorescence is preserved (see, e.g., references in de Boer et al. [1]). Serial sections are then loaded into the integrated microscope and imaged sequentially. Automated procedures for registration between imaging modalities can be implemented to ensure consistent overlay accuracy across an entire reconstructed dataset [14].

Future applications of CLEM will demand greater precision, further automation, and higher throughput, for which iCAT offers a number of potential advantages. Above all else, iCAT enables large numbers of serial sections to be sequentially and automatically imaged to generate reconstructed volumes of overlaid FM and EM datasets with matching axial resolution. Moreover, specimen warping and/or shrinkage, which might otherwise occur in conventional array tomography methods, is prevented due to the absence of intermediate sample preparation. This ensures a precise overlay of biological molecules and structural context at high resolution in all three dimensions. Additionally, precisely overlaid fluorescence data has the potential to vastly improve classification of ultrastructural features in EM data.

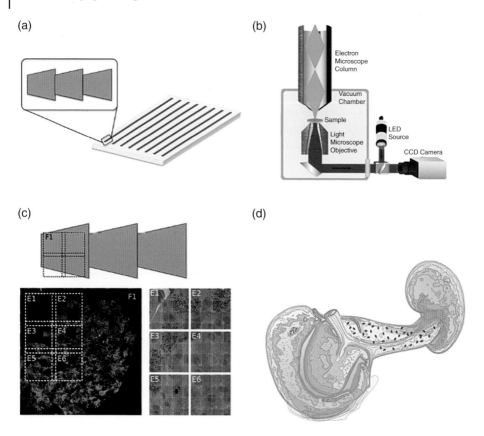

Figure 7.1 Conceptual overview of a workflow for integrated Correlative Array Tomography (iCAT). (a) Serial sections are collected in ribbons and mounted on a glass slide coated with transparent, conductive indium tinoxide. (b) This slide is then inserted into the integrated microscope. (c) A series of FM and EM images are then acquired according to an automated imaging scheme. FM image tiles of each tissue section are acquired prior to EM imaging to avoid electron beam bleaching. Following FM acquisition, high-resolution EM images are acquired over regions identified by fluorescence expression. (d) Images are post-processed to create a 3D CLEM reconstruction where FM and EM data are precisely overlaid.

As alluded to previously, while modern machine-learning-based segmentation methods (e.g. ilastik [15], SuRVoS [16]) are quite sophisticated, they nevertheless require some degree of manual annotation. Because high-accuracy overlaid correlative datasets contain, in a sense, the classification data that these methods seek to provide, such datasets could reduce the need for supervised learning while opening up new possibilities for different machine learning applications. For combating throughput limitations, iCAT is equipped to enable automated detection of regions of interest based on in-section fluorescence expression. And by sharing a common optical axis and translation stage, it is possible to drive to the select cell or region within the section to begin high-resolution imaging. This kind of automated imaging scheme has already seen early development and implementation [17]. Finally, fluorescence preservation or post-embedding re-labeling of genetic fluorophores may facilitate linking ultrastructural observations to live, intravital fluorescence microscopy. Prior to array tomography, photoactivatable probes could be used to mark where cells can be assessed for function. These markers could then be activated as part of the iCAT workflow, thus linking the ultrastructural data to not only fluorescence expression but also function and development [6].

7.3 Super-Resolution Fluorescence Microscopy

7.3.1 Advantages and Challenges for CLEM with Super-Resolution Fluorescence

In many CLEM approaches, including some of the above, fluorescence microscopy (FM) serves mostly as a tool to rapidly screen for regions of interest which are then subjected to EM for further structural analysis. In those approaches, diffraction-limited FM is sufficient for successful EM-FM correlation. However, in cases where the biological question requires precise molecular localization within the EM ultrastructure, it is necessary to turn to super-resolution (SR) fluorescence to circumvent the diffraction limit of light. Alternatively, CLEM with SR-FM (SR-CLEM) can reveal formerly concealed information [18–20]. Further, complementary to the apparent advantages in SR-CLEM, such advents can even act as fidelity-checks for SR results, thereby generating highly reliable, correlated and possibly even quantitative datasets [21]. Near EM-resolution in FM data, when perfectly registered can "colorize" EM with true molecular context.

With improved FM resolution, however, the challenges in conventional diffraction limited CLEM are taken a notch higher. First, there is less tolerance to registration inaccuracies. Especially for SR-CLEM (with optical resolutions of 10–20 nm, approaching that of EM), strictly SR-FM to EM registration accuracies comparable to or better than the super-resolution values are required. Further, there is increased demand on sample preparation maintaining brightness and photo-stability of (potentially special) fluorophores, and minimal to no background fluorescence, e.g. from embedding media used in EM. SR-FM in sequential CLEM has been demonstrated albeit with stringent constraints [22–24]. Registration between FM and EM can be demanding and typically involves, if discernible in both systems, (manual) feature identification and/or fiducial markers. In diffraction-limited FM, manual registration may well be sufficient [25], especially with conspicuous structures such as the mitochondria. Finder grids have also been employed [26] and are comparable to manual registration in their accuracies. Fiducial markers, such as nanoparticles, can provide registration accuracy well below 100 nm [27, 28], possibly even down to 10 nm, thus comparable to the resolution obtainable with SR-FM and suitable for SR-CLEM [24]. However, this carries the risk of not having sufficient numbers of fiducials in the ROI to perform the correlation at the aimed-for accuracy. Also, when the fiducials are not embedded in the sample but for instance in the sample support, sample distortions occurring during intermediate sample preparation may unnoticeably compromise the registration. Hence, for SR-CLEM to deliver its full potential, it remains a challenge to consistently obtain highly accurate correlation throughout a sample without errors due to registration or post-FM sample deformation.

7.3.2   Implementation of SR-FM with CLEM

Integrated microscopes can remove the need for external/manual alignment markers by offering automated registration procedures based on electron-induced signals detectable in the FM [14]. Furthermore, high registration accuracies (down to ~5 nm) are not only achievable, but consistent across the sample/region of interest [14]. Naturally, inspection with an integrated microscope is devoid of intermediate sample deformations, offering a seemingly unique platform for SR-CLEM. However, this also necessitates samples that are simultaneously amenable to both FM and EM and maintain their SR properties after fixation, plastic-embedding, and staining and in the vacuum environment of the integrated microscope. For SR-CLEM in sequential approaches protocols have been developed to preserve the fluorescence via weaker EM fixation and diluted staining at the expense of ultrastructural preservation [23, 29].

Fluorescence in standard fluorescence proteins (FPs) such as the green fluorescent protein (GFP) and mCherry can also be preserved in-resin using optimized sample preparation [30]. Engineering FPs that are characteristically more robust to osmium fixation and uranyl or other heavy metal staining, like the recent mEos4, [31] opens the doors for SR-CLEM without compromise. However, while synthetic fluorophores are quite capable of surviving in vacuum [32], FPs are more severely dependent on hydrated and oxygen-rich environments. Recently, in-resin SR-CLEM on FPs in an integrated setup has been demonstrated using an environmental chamber in which a water vapor can be introduced. These results even indicated that standard FPs like GFP and yellow fluorescent protein (YFP) show better blinking characteristics for SR-FM under the partial vacuum induced by injecting water vapor than in ambient conditions [20]. Both these avenues, engineered fixation-resistant FPs and environmental in-resin standard FPs, hold great promise to meet the challenges in SR-CLEM, but also call for further development to a wider palette of fluorescent molecules and preparation protocols. For instance, while mEos4 is successful in hydrophilic resins like the GMA, HM20 and variants, efforts are underway in translating to epoxy based resins like epon, which may be preferred for some applications [31].

7.3.3   Prospects for Integrated SR-CLEM

With these initial demonstrations and further development en-route, we may wonder what innovations might be next and what further advantages could integrated microscopy offer particularly for SR-CLEM? First, it has to be realized that SR-FM is intrinsically a low-throughput method. Often, a single SR image may take several minutes to be acquired, and often extensive post-processing is also required. The characteristic large FOV of a wide-field FM image may subsequently be reduced to a small ROI, also due to the high NA objectives which are mandatory for SR-FM. Pre-screening with standard diffraction limited FM is usually a prerequisite [21]. When performed in a sequential fashion, these time-overheads easily add up, especially when additional processing steps such as fixation, staining, sample movement and ROI retrieval are involved. Integrated systems can reduce the time costs and inherently lend themselves towards automation and higher throughputs. Pre-programmed widefield FM detection, in combination with array tomography, could also be viable, with SR-FM being performed only in dedicated regions.

In addition, integrated systems, in which the sample can be simultaneously imaged by EM and FM, have the potential to add to the palette of available SR techniques. Especially for dyes where vacuum or other constraints preclude conventional SR-FM, the availability of simultaneous multi-modal information and the ability to control the experimental system in a central fashion with both light and electrons may offer innovative ways of manipulating, processing, and interpreting fluorescence data to achieve SR. Direct electron excitation, or cathodoluminescence [33, 34], has been explored but is impossible for organic molecules due to rapid molecular degradation, or bleaching, under electron irradiation [35]. An alternative approach may be to use the electron-beam to selectively bleach or otherwise modify the fluorescence signal. By correlating this instantaneous drop in intensity with the concomitant SEM signals, an SR image can be reconstructed with perfect registration to EM – enabling a resolution that is ultimately determined by the electron-interaction volume. The reduced dependence on unique fluorophore properties (most of them bleach) is an added advantage that offers broader application options.

Undeniably, the investment in integrated workflows may not be adequately rewarding for all users. However, for samples that require large-volumes to be diligently inspected, serial sectioning and resin-embedding offer the largest integrity in ultrastructure preservation and sectioning losses [31]. It is also these that would highly benefit from automated workflows, ROI detection with fluorescence and ultimately, limited expert human involvement. These points are in accordance with what an integrated system requires, already has, and/or can eventually provide. The capability of on-demand SR-FM in such a system thus makes for a more complete, unified imaging platform.

7.4 Cryo-Electron Microscopy

7.4.1 Advantages of CryoEM

With cryo-electron microscopy (cryoEM) vitrified samples are imaged in the electron microscope under cryogenic conditions. A unique benefit of cryoEM over room-temperature EM is the structural preservation of the sample at molecular level, as a result of sample freeze fixation by vitrification. This excellent structural preservation, in combination with the use of 3D image acquisition and processing strategies, enables the 3D reconstructions of cells, bacteria, viruses, and biomolecules. Resolution limitations are set by intrinsic electron-beam-induced radiation damage to the specimen during imaging. During the last decade, a series of technological improvements [36] made it possible to image molecular structures to near-atomic resolution by combining the images of a large number of virtually identical structures (particles), either from many 2D views (single particle analysis, SPA) [37], or 3D volumes from 3D tomograms (sub-tomogram averaging, STA) [38]. The near-atomic resolution reconstructions by SPA are typically the result of averaging identical and repetitive structures in a suspension of purified molecules. For targeting, imaging, reconstructing and interpreting inherently flexible structures in the cellular context of the cell, cryo-electron tomography (cryoET) is the only available technique to obtain a 3D reconstruction of the molecular arrangement. The attainable resolution of cryoET is less than the results obtained by single-particle approaches, as the averaging strategy to enhance the signal-to-noise ratio uses multiple exposures at different angles of the same position. Nevertheless, using sub-tomogram averaging strategies in which multiple copies of the same structure are averaged from tomograms, impressive and encouraging results by cryoET of imaging molecular structures have been obtained [38]. One of the challenges of high-resolution cellular cryoET is the prerequisite that the thickness dimension should be less than a few 100 nm. So, a workflow that includes a way to physically thin a vitrified (cellular) volume using a focused ion beam (FIB) to mill away large parts of cells into a thin lamella, is transformative to the field [39]. However, FIB milling currently has a limited availability and throughput, is highly demanding in terms of time, equipment, and expertise, and is not a routine procedure in the field.

7.4.2 Possibilities and Challenges for Correlative Cryo-Microscopy

One of the essential requirements in performing structural biology studies in the context of the cell is the capability to target specific structures. The use of antibody-coupled electron dense gold labels as are often used in EM on sections at room temperature cannot be generally used for labeling in live cells, though several attempts have been made [40]. Therefore, the use of genetic or chemical types of labeling to introduce a fluorescent light microscopy marker in cells that is suitable to identify and locate a specific molecular structure at molecular (nm-scale) spatial resolution in the electron microscope would be convenient. During the last decades, FM achieved broad application within cell biology and a plethora of labels are available for live cell imaging. Building upon these developments, probes for live-cell FM can also be used for imaging in cryogenic samples like vitrified cells using cryo light microscopy stages [41–43]. This combination of FM imaging under cryogenic conditions with cryoET in a correlative workflow can be very powerful as it harbors the possibility of identifying a structure based on its specific spectral signature and target the three-dimensional spatial location of fluorescently labeled molecules inside biological samples. Therefore, cryo-correlative light and electron microscopy (cryoCLEM) could encompass the grail of targeting and imaging specific single molecules in the interior a cell.

While several cryoCLEM workflows have been established [44–46] and it seems that in theory cryoCLEM could provide a way to perform structural biology in the context of the cell, in practice however, there remain considerable challenges to achieving molecular resolution tomographic imaging of individually targeted macromolecules, and several technical and theoretical issues are not yet fully comprehended. In order to answer a particular structural cell biological question, it is crucial that these individual steps are well understood and combined into efficient and robust workflows. So what steps are required, and which need more extensive improvement and attention from the field?

7.4.2.1 Super-Resolution Fluorescence Cryo-Microscopy: Probes and Instruments

Single molecule or SR cryoFM is necessary to attain localization of a molecular structure within the vitrified biological samples. The resolving power of diffraction limited light microscopy (±200 nm) is not suitable to localize individual proteins (± 5 nm) inside the crowded environment within the cell. Nondiffraction limited SR-FM can achieve a resolving power close to the size of proteins. Several groups are exploring SR imaging of vitrified specimen at cryogenic conditions [47–49], but to date with limited application.

Several issues play a role that make SR-cryoFM a challenge. First, the stability of cryogenic light microscopy stages needs to be significant during the relatively long exposure times that are necessary during SR-FM imaging. With necessary imaging times (~30 minutes) and desirable resolutions (~10 nm), not only the lateral (X and Y) shifts, which can be corrected for by use of fiducial markers such as fluorescent nanoparticles, but especially the axial (height or Z) shifts should be limited. These stabilities are difficult to achieve at cryo conditions due to constant heat transfer from the surrounding and the objective lens being in close proximity to the specimen. In addition, depending on the technical implementation of the cryo-light microscope, a long-working lens may be necessary to limit heat transfer, which may come at the expense of the numerical aperture, resulting in less efficient photon counting and thus longer imaging times and possibly lower localization accuracy. Second, probes suitable for SR-FM are developed, tested, and optimized for (live cell) imaging at ambient temperatures. The performance and applicability under cryogenic conditions of those probes and especially for specific forms of non-diffraction limited or localization based on SR-imaging techniques might not be suitable or just not work, and specific development, testing and optimization of probes working under cryogenic conditions is necessary. For example, it is likely that the switching, activation, and deactivation of many probes and dyes are hampered at cryogenic conditions. In addition, for several of the SR-FM techniques, a prolonged illumination with light or illumination with multiple wavelengths is required that could lead to damage, e.g. due to heating of the sample and consequently devitrification.

7.4.2.2 Transfer of Cryo-Samples between Microscopes

Following cryoFM, the same sample needs to be inspected with cryoEM. In a two-step cryoCLEM approach, the sample is transferred from the light to the electron microscope under cryogenic conditions. As with CLEM with resin embedding at room temperature (see also 1.4.1), this step requires a relocation accuracy between instruments in the same order of magnitude as the desired final resolution. In current two-step implementations, this can to a large extent be carried out using additional bimodal fiducial markers, electron-dense and fluorescent markers like quantum dots or fluorescent beads that are detectable both in light and electron microscopy, which can be used for accurate navigation to the fluorescent structure of interest by means of spatial coordinate interpolation [24, 25]. Apart from this, in practice, the transfer of cryogenically cooled grids between instruments is even more time consuming and prone to sample degradation than for CLEM with samples at room temperature due to manual handling and humidity contamination from the air. Both carrier systems and software for relocation and overlay of images are indispensable for fast and accurate relocation and image overlay for later presentation.

7.4.2.3 Sample Thickness

During the past few years, several commercial realizations of instruments to physically dissect a volume or lamella from a vitrified biological structure have become available (e.g., Thermo-Fisher Scientific Aquilos and Zeiss Crossbeam). While in materials science thinning the sample by FIB-milling has been a standard technique for some time, only in last years has the technique also been adapted for biological tissue. For the purpose of cryoCLEM, the goal of this step is to dissect a lamella with a thickness in the order of a few 100 nm that contains the labeled structure as identified during the previous cryoFM step. The current performance and technical availability of this step in the workflow shows that indeed it is feasibly to dissect such lamella from vitrified cells. This has revolutionized the possibilities of performing structural biology inside cells, since these lamellae provide an unperturbed view at the interior of a cell [39]. Several issues need to be overcome in order to lift the current state-of-the-art technique to a more routine EM method.

One fundamental limitation is the quality of intracellular vitrification variability prior to the FIB-milling. Vitrification depths that can be achieved with plunge-freezing, without the formation of crystalline ice, are in the range of 10 microns, resulting in large, significant volumes of cells, which are thicker, not being appropriately freeze-fixed. Methods for high-pressure-freezing on EM grids have been developed to overcome this [50]. Furthermore, the current success rate of creating stable (unbroken) FIB-milled lamellae is not high (~50%) due to technical limitations (e.g. sample drift, contamination within the instrument).

Another technical limitation is the capability to dissect a lamella from the cell containing the fluorescently labeled structure. The required 3D location needs to be such that the structure is located within the lamella, therefore having an accuracy on the order of 50 nm. Existing approaches based on performing confocal light cryo-microscopy on the vitrified sample prior to milling have demonstrated an accuracy on the order of 500 nm [51]. Nevertheless, it seems likely that with focused technical attention, this challenge can be met. A workaround to this limitation is to acquire a large number of lamellae and, subsequently, image those lamellae with high-resolution cryoFM to select those that contain the structure of interest. Clearly, there are several obstacles to carrying out this workaround in practice due to the low throughput and additional transfer step from the microscope for cryoFIB-to the cryoFM microscope.

7.4.2.4 Data Collection Speed

For near-atomic resolution imaging of structures by cryoET, many structures have to be targeted, imaged and averaged. On purified, reconstituted or concentrated biological samples many copies will be present in a few images or tomograms. In a cell, however, structures might not be present in many copies. Therefore, to obtain a sufficient amount of data for structure determination, a large number of images/tomograms might be necessary. Note that for non-repetitive structures, a large amount of electron tomograms is not only needed for the (mathematical) averaging, but also required to make biological-relevant statements related to number of occurrences of the specific molecules in a particular biological setting. To provide biological relevant observations suitable for quantitative analyses, data collection speed should be increased by at least a factor of ten in order to have sufficient number of particles (104) within a reasonable amount of time (days). In current setups, the speed of data collection is in the order of an hour per tomogram, and this is limited by camera read-out and data transfer speeds and stage stability and accuracy. Improvements are possible, since fast and accurate stages are available for room temperature data collection and using material science samples it has been possible to more rapidly acquire atomic resolution structures [52]. So, while in principle the electronic read-out speed of state-of-the-art digital cameras is sufficient for fast data collection, current movie-frame recording is only used for post-imaging alignments and high-resolution reconstructions [53] and not yet for fast recording of tilt series, which is necessary for significantly faster data collection. Faster data collection would also push the speed of subsequent data storage and tomographic tilt series alignment and reconstruction algorithms. Implementations of live alignment and reconstruction algorithms [54] will be necessary to achieve the goal of live feedback on data collection.

7.4.3 Integrated Systems for CryoCLEM

Having looked at some of the individual steps in the cryoCLEM workflow, it becomes clear that each of these steps has, during the last few years, been shown capable of producing highly valuable results. Nevertheless, the execution of a full cryoCLEM workflow is, in practice, a state-of-the-art technique that has been executed only in a few cases [51, 55–58]. One of the main limitations is the necessity to transfer the sample between multiple instruments (e.g. FIB-SEM to FM to TEM), which is both time-consuming, and prone to low yield, carrying the risk of degradation of sample quality, and being prone to transfer errors. An integrated system, combining the functionality of sample thinning, FM, and TEM into one instrument, would not only mitigate the risks related to transfer from one imaging modality to the other, but could also, to a large extent, lead to automated execution of the individual steps. Automation would ultimately result in an instrument that is simpler to use and delivers more robust and most likely faster performance with a higher relocation accuracy.

The integrated systems that were created for cryoCLEM during the past few years were mainly prototypes. So far, they have covered a limited amount of the aspects of the full cryoCLEM workflow. Examples of such integrated systems are: a light microscope integrated in a TEM [59], and a light microscope built in a plunging device prior to vitrification [60]. While each of these solutions has shown its merits, they did not provide an overall better solution compared to the workflow composed of the individual instruments, mainly due to the technical limitations of these integrated systems, e.g. affecting quality or resolution and/or versatility and flexibility. Nevertheless, a number of lessons can be learned from these integrated systems to attempt to create better integrated solutions. Results on the behaviour and characteristics of fluorophores under cryogenic conditions, the effects of interference in vitrified cells when imaging with a one-wavelength laser, the effects of the illumination on a specimen in a TEM that results in more heating than expected, all clarify the limitations and opportunities of integrated cryoCLEM systems.

7.4.4 Prospects for Integrated Cryo-Microscopy

We envision that the current cryoCLEM workflow, which is based on sequential use of individual instruments, will mature in the coming years resulting in a tool providing unprecedented results for cell and structural biology. On the short term there will primarily be an increase in performance of current individual systems: (i) improved stages and microscopes for cryoFM, including SR-FM; (ii) better technological solutions to specimen shuttling; (iii) more reliable performance of cryoFIB instruments; and (iv) faster cryoET. These are currently present and/or under development and improvements will become available in the near future. The flexibility to adapt the optics, mechanics, automation, electronics, and more of each of the individual steps is enormous and will provide the knowledge to create a reliable combination of these instruments. Therefore, as the knowledge related to cryoCLEM with the different modalities will increase over time, we anticipate that on the long-term, technological solutions that cover parts of the workflow will also become available.

Over the long term, this increased knowledge related to the individual steps involved in cryoCLEM must be used to enforce more drastic improvements in technology. The main problem with integrated systems is the inefficient use and limited capabilities that are posed by the combination of different techniques. For example, the integrated light and electron microscope for cryoEM use was limited by the type of microscope it was built in. The Tecnai 120 keV microscope with LaB_6 source, twin lens, and standard camera does not compare to a 300 keV TEM with FEG source and direct electron detector camera for near-atomic resolution imaging by cryoEM or cryoET. On the other hand, these high-end microscopes are designed from scratch for high-throughput cryoEM and not for correlative imaging. So newly designed, designated cryoCLEM systems for cellular imaging are necessary to advance cryoCLEM, leading to fully integrated solutions that are optimized for specific applications and that, for those applications, are more suitable than a combination of optimized instruments. In the case of cryoCLEM this ultimately means fixation, cryo-SR-FM, cryo-FIB and TEM tomography in one integrated solution. Partial solutions that integrate part of the entire workflow without compromising on the performance of stand-alone microscopes, paired with standardized and/or automated transfer steps may constitute the first steps toward this final goal.

7.5   Outlook

In the preceding paragraphs, we have discussed three of the grand challenges in EM and CLEM and presented our view on how integrated microscopy may contribute to the road ahead. Prospects for each of the areas of large-scale and large-volume EM, super-resolution FM with EM, and cryoEM are given at the end of the respective sections. Key to several of the presented solutions is automating the workflow and reducing the number of human (manual) interventions. Clearly, as stated before, setting up an integrated workflow will not be a practical solution for each individual experiment. But on the other hand, throughput limitations currently are the main bottleneck in almost all major microscopy facilities, which is mostly due to the human labor involved in the experimental workflow. In this respect, we may well look at the field of electron beam lithography and cleanroom nanofabrication, where workflows for recurring tasks can be operated fully automated with programmable machines in a contamination-free environment, while only specific jobs (e.g. for exploratory research) are still going through a dedicated workflow involving an expert technician or operator. Extending this approach to EM, we would need identification of recurring tasks/questions for which standardized sample preparation protocols could be developed. We could envision a wafer tray next to the microscope onto which samples can be loaded while the machine is running 24/7. Samples are loaded automatically from the wafer tray into the microscope together with a pre-programmed set of instructions for recognition of regions of interest (e.g. based on expressed fluorescence). Thus, the microscope operates independently, controls its settings and performance, applies the needed (integrated) modalities, and provides the end user not just with operator-selected, representative scientific images, but with scientific, possibly even quantified, statistical, data.

Acknowledgments

We thank Pascal de Boer and Ben Giepmans (UMC Groningen) for the sample shown in Figure 7.1c. We thank our lab members and collaborators for the many fruitful discussions that shaped the views displayed in this chapter. We acknowledge financial support from the Netherlands Organisation for Scientific Research through the NWO-TTW Perspectief program Microscopy Valley (projects 12713, 12714), NWO-TTW Open Technologie project 15313 (A. S. R.), and NWO-BBoL grant 737.016.010 (R. I. L.).

Statement of financial interest: J. P. H. has a financial interest in Delmic B.V., a company providing equipment for integrated microscopy.

References

1 de Boer, P., Hoogenboom, J.P. and Giepmans, B.N.G. Correlated light and electron microscopy: ultrastructure lights up! *Nature Methods* **12**, 503–513 (2015).

2 Zonnevylle, A.C., Van Tol, R.F.C., Liv, N., Narvaez, A.C., Effting, A.P.J., Kruit, P. and Hoogenboom, J.P. Integration of a high-NA light microscope in a scanning electron microscope. *Journal of Microscopy* **252**, 58–70 (2013).

3 Timmermans, F.J. and Otto, C. Review of integrated correlative light and electron microscopy. *Review of Scientific Instruments* **86** (2015).

4 Briggman, K.L. and Bock, D.D. Volume electron microscopy for neuronal circuit reconstruction. *Current Opinion in Neurobiology* **22**, 154–161 (2012).

5 Peddie, C.J. and Collinson, L.M. Exploring the third dimension: Volume electron microscopy comes of age. *Micron* **61**, 9–19 (2014).

6 Collinson, L.M., Carroll, E.C. and Hoogenboom, J.P. Correlating 3D light to 3D electron microscopy for systems biology. *Current Opinion in Biomedical Engineering* **3**, 49–55 (2017).

7 Hildebrand, D.G.C., Cicconet, M., Torres, R.M., Choi, W., Quan, T.M., Moon, J., Wetzel, A.W., Champion, A.S., Graham, B.J., Randlett, O., Plummer, G.S., Portugues, R., Bianco, I.H., Saalfeld, S., Baden, A.D., Lillaney, K., Burns, R., Vogelstein, J.T., Schier, A.F., Lee, W.-C.A., Jeong, W.-K., Lichtman, J.W. and Engert, F. Whole-brain serial-section electron microscopy in larval zebrafish. *Nature* **545**, 345–349 (2017).

8 Bock, D.D., Lee, W.C.A., Kerlin, A.M., Andermann, M.L., Hood, G., Wetzel, A.W., Yurgenson, S., Soucy, E.R., Kim, H.S. and Reid, R.C. Network anatomy and in vivo physiology of visual cortical neurons. *Nature* **471**, 177–U159 (2011).

9 Zheng, Z., Lauritzen, J.S., Perlman, E., Robinson, C.G., Nichols, M., Milkie, D., Torrens, O., Price, J., Fisher, C.B., Sharifi, N., Calle-Schuler, S.A., Kmecova, L., Ali, I.J., Karsh, B., Trautman, E.T., Bogovic, J., Hanslovsky, P., Jefferis, G.S.X.E., Kazhdan, M., Khairy, K., Saalfeld, S., Fetter, R.D. and Bock, D.D. A Complete Electron Microscopy Volume Of The Brain Of Adult Drosophila melanogaster. *bioRxiv* (2017).

10 Karreman, M.A., Mercier, L., Schieber, N.L., Solecki, G., Allio, G., Winkler, F., Ruthensteiner, B., Goetz, J.G. and Schwab, Y. Fast and precise targeting of single tumor cells in vivo by multimodal correlative microscopy. *J Cell Sci* **129**, 444–456 (2016).

11 Karreman, M.A., Hyenne, V., Schwab, Y. and Goetz, J.G. Intravital Correlative Microscopy: Imaging Life at the Nanoscale. *Trends Cell Biol* **26**, 848–863 (2016).

12 Micheva, K.D. and Smith, S.J. Array tomography: A new tool for Imaging the molecular architecture and ultrastructure of neural circuits. *Neuron* **55**, 25–36 (2007).

13 Wacker, I. and Schroeder, R.R. Array tomography. *Journal of Microscopy* **252**, 93–99 (2013).

14 Haring, M.T., Liv, N., Zonnevylle, A.C., Narvaez, A.C., Voortman, L.M., Kruit, P. and Hoogenboom, J.P. Automated sub-5 nm image registration in integrated correlative fluorescence and electron microscopy using cathodoluminescence pointers. *Scientific Reports* **7** (2017).

15 Sommer, C., Strähle, C., Köthe, U. and Hamprecht, F.A. in Eighth IEEE International Symposium on Biomedical Imaging (ISBI) 230–233 (2011).

16 Luengo, I., Darrow, M.C., Spink, M.C., Sun, Y., Dai, W., He, C.Y., Chiu, W., Pridmore, T., Ashton, A.W., Duke, E.M.H., Basham, M. and French, A.P. SuRVoS: Super-Region Volume Segmentation workbench. *Journal of Structural Biology* **198**, 43–53 (2017).

17 Delpiano, J., Pizarro, L., Peddie, C.J., Jones, M.L., Griffin, L.D. and Collinson, L.M. Automated detection of fluorescent cells in in-resin fluorescence sections for integrated light and electron microscopy. *Journal of Microscopy* (2017).

18 Kopek, B.G., Shtengel, G., Xu, C.S., Clayton, D.A. and Hess, H.F. Correlative 3D superresolution fluorescence and electron microscopy reveal the relationship of mitochondrial nucleoids to membranes. *Proceedings of the National Academy of Sciences of the United States of America* **109**, 6136–6141 (2012).

19 Loschberger, A., Franke, C., Krohne, G., van de Linde, S. and Sauer, M. Correlative super-resolution fluorescence and electron microscopy of the nuclear pore complex with molecular resolution. *Journal of Cell Science* **127**, 4351–4355 (2014).

20 Peddie, C.J., Domart, M.C., Snetkov, X., O'Toole, P., Larijani, B., Way, M., Cox, S. and Collinson, L.M. Correlative super-resolution fluorescence and electron microscopy using conventional fluorescent proteins in vacuo. *J Struct Biol* **199**, 120–131 (2017).

21 Hauser, M., Wojcik, M., Kim, D., Mahmoudi, M., Li, W. and Xu, K. Correlative Super-Resolution Microscopy: New Dimensions and New Opportunities. *Chemical Reviews* **117**, 7428–7456 (2017).

22 Betzig, E., Patterson, G.H., Sougrat, R., Lindwasser, O.W., Olenych, S., Bonifacino, J.S., Davidson, M.W., Lippincott-Schwartz, J. and Hess, H.F. Imaging intracellular fluorescent proteins at nanometer resolution. *Science* **313**, 1642–1645 (2006).

23 Watanabe, S., Punge, A., Hollopeter, G., Willig, K.I., Hobson, R.J., Davis, M.W., Hell, S.W. and Jorgensen, E.M. Protein localization in electron micrographs using fluorescence nanoscopy. *Nature Methods* **8**, 80–U117 (2011).

24 Kopek, B.G., Shtengel, G., Grimm, J.B., Clayton, D.A. and Hess, H.F. Correlative photoactivated localization and scanning electron microscopy. *PLoS One* **8**, e77209 (2013).

25 Sjollema, K.A., Schnell, U., Kuipers, J., Kalicharan, R. and Giepmans, B.N.G. in Correlative Light and Electron Microscopy, Vol. 111. (eds. T. MullerReichert and P. Verkade) 157–173 (2012).

26 Spiegelhalter, C., Tosch, V., Hentsch, D., Koch, M., Kessler, P., Schwab, Y. and Laporte, J. From Dynamic Live Cell Imaging to 3D Ultrastructure: Novel Integrated Methods for High Pressure Freezing and Correlative Light-Electron Microscopy. *Plos One* **5** (2010).

27 Kukulski, W., Schorb, M., Welsch, S., Picco, A., Kaksonen, M. and Briggs, J.A.G. Correlated fluorescence and 3D electron microscopy with high sensitivity and spatial precision. *Journal of Cell Biology* **192**, 111–119 (2011).

28 Schellenberger, P., Kaufmann, R., Siebert, C.A., Hagen, C., Wodrich, H. and Grunewald, K. High-precision correlative fluorescence and electron cryo microscopy using two independent alignment markers. *Ultramicroscopy* **143**, 41–51 (2014).

29 Kopek, B.G., Paez-Segala, M.G., Shtengel, G., Sochacki, K.A., Sun, M.G., Wang, Y.L., Xu, C.S., van Engelenburg, S.B., Taraska, J.W., Looger, L.L. and Hess, H.F. Diverse protocols for correlative super-resolution fluorescence imaging and electron microscopy of chemically fixed samples. *Nature Protocols* **12**, 916–946 (2017).

30 Peddie, C.J., Blight, K., Wilson, E., Melia, C., Marrison, J., Carzaniga, R., Domart, M.C., O'Toole, P., Larijani, B. and Collinson, L.M. Correlative and integrated light and electron microscopy of in-resin GFP fluorescence, used to localise diacylglycerol in mammalian cells. *Ultramicroscopy* **143**, 3–14 (2014).

31 Paez-Segala, M.G., Sun, M.G., Shtengel, G., Viswanathan, S., Baird, M.A., Macklin, J.J., Patel, R., Allen, J.R., Howe, E.S., Piszczek, G., Hess, H.F., Davidson, M.W., Wang, Y. and Looger, L.L. Fixation-resistant photoactivatable fluorescent proteins for CLEM. *Nature Methods* **12**, 215–+ (2015).

32 Karreman, M.A., Agronskaia, A.V., van Donselaar, E.G., Vocking, K., Fereidouni, F., Humbel, B.M., Verrips, C.T., Verkleij, A.J. and Gerritsen, H.C. Optimizing immuno-labeling for correlative fluorescence and electron microscopy on a single specimen. *Journal of Structural Biology* **180**, 382–386 (2012).

33 Glenn, D.R., Zhang, H., Kasthuri, N., Schalek, R., Lo, P.K., Trifonov, A.S., Park, H., Lichtman, J.W. and Walsworth, R.L. Correlative light and electron microscopy using cathodoluminescence from nanoparticles with distinguishable colours. *Scientific Reports* **2** (2012).

34 Garming, M.W.H., Weppelman, I.G.C., de Boer, P., Perona Martinez, F., Schirhagl, R., Hoogenboom, J.P. and Moerland, R.J. Nanoparticle discrimination based on wavelength and lifetime-multiplexed cathodoluminescence microscopy. *Nanoscale* **9**, 12727–12734 (2017).

35 Niitsuma, J., Oikawa, H., Kimura, E., Ushiki, T. and Sekiguchi, T. Cathodoluminescence investigation of organic materials. *Journal of Electron Microscopy* **54**, 325–330 (2005).

36 Koning, R.I., Koster, A.J. and Sharp, T.H. Advances in cryo-electron tomography for biology and medicine. *Annals of Anatomy* **217**, 82–96 (2018).

37 Kuhlbrandt, W. Biochemistry. The resolution revolution. *Science* **343**, 1443–1444 (2014).

38 Schur, F.K., Obr, M., Hagen, W.J., Wan, W., Jakobi, A.J., Kirkpatrick, J.M., Sachse, C., Krausslich, H.G. and Briggs, J.A. An atomic model of HIV-1 capsid-SP1 reveals structures regulating assembly and maturation. *Science* **353**, 506–508 (2016).

39 Mahamid, J., Pfeffer, S., Schaffer, M., Villa, E., Danev, R., Cuellar, L.K., Forster, F., Hyman, A.A., Plitzko, J.M. and Baumeister, W. Visualizing the molecular sociology at the HeLa cell nuclear periphery. *Science* **351**, 969–972 (2016).

40 Diestra, E., Fontana, J., Guichard, P., Marco, S. and Risco, C. Visualization of proteins in intact cells with a clonable tag for electron microscopy. *Journal of Structural Biology* **165**, 157–168 (2009).

41 Sartori, A., Gatz, R., Beck, F., Rigort, A., Baumeister, W. and Plitzko, J.M. Correlative microscopy: bridging the gap between fluorescence light microscopy and cryo-electron tomography. *Journal of Structural Biology* **160**, 135–145 (2007).

42 Schwartz, C.L., Sarbash, V.I., Ataullakhanov, F.I., McIntosh, J.R. and Nicastro, D. Cryo-fluorescence microscopy facilitates correlations between light and cryo-electron microscopy and reduces the rate of photobleaching. *Journal of Microscopy* **227**, 98–109 (2007).

43 van Driel, L.F., Valentijn, J.A., Valentijn, K.M., Koning, R.I. and Koster, A.J. Tools for correlative cryo-fluorescence microscopy and cryo-electron tomography applied to whole mitochondria in human endothelial cells. *European Journal of Cell Biology* **88**, 669–684 (2009).

44 Hampton, C.M., Strauss, J.D., Ke, Z., Dillard, R.S., Hammonds, J.E., Alonas, E., Desai, T.M., Marin, M., Storms, R.E., Leon, F., Melikyan, G.B., Santangelo, P.J., Spearman, P.W. and Wright, E.R. Correlated fluorescence microscopy and cryo-electron tomography of virus-infected or transfected mammalian cells. *Nature Protocols* **12**, 150–167 (2017).

45 Koning, R.I., Celler, K., Willemse, J., Bos, E., van Wezel, G.P. and Koster, A.J. Correlative cryo-fluorescence light microscopy and cryo-electron tomography of Streptomyces. *Methods in Cell Biology* **124**, 217–239 (2014).

46 Schorb, M., Gaechter, L., Avinoam, O., Sieckmann, F., Clarke, M., Bebeacua, C., Bykov, Y.S., Sonnen, A.F., Lihl, R. and Briggs, J.A.G. New hardware and workflows for semi-automated correlative cryo-fluorescence and cryo-electron microscopy/tomography. *Journal of Structural Biology* **197**, 83–93 (2017).

47 Chang, Y.W., Chen, S., Tocheva, E.I., Treuner-Lange, A., Lobach, S., Sogaard-Andersen, L. and Jensen, G.J. Correlated cryogenic photoactivated localization microscopy and cryo-electron tomography. *Nature Methods* **11**, 737–739 (2014).

48 Liu, B., Xue, Y., Zhao, W., Chen, Y., Fan, C., Gu, L., Zhang, Y., Zhang, X., Sun, L., Huang, X., Ding, W., Sun, F., Ji, W. and Xu, T. Three-dimensional super-resolution protein localization correlated with vitrified cellular context. *Scientific Reports* **5**, 13017 (2015).

49 Kaufmann, R., Schellenberger, P., Seiradake, E., Dobbie, I.M., Jones, E.Y., Davis, I., Hagen, C. and Grunewald, K. Super-resolution microscopy using standard fluorescent proteins in intact cells under cryo-conditions. *Nano Lett* **14**, 4171–4175 (2014).

50 Harapin, J., Bormel, M., Sapra, K.T., Brunner, D., Kaech, A. and Medalia, O. Structural analysis of multicellular organisms with cryo-electron tomography. *Nature Methods* **12**, 634–636 (2015).

51 Arnold, J., Mahamid, J., Lucic, V., de Marco, A., Fernandez, J.J., Laugks, T., Mayer, T., Hyman, A.A., Baumeister, W. and Plitzko, J.M. Site-Specific Cryo-focused Ion Beam Sample Preparation Guided by 3D Correlative Microscopy. *Biophysical Journal* **110**, 860–869 (2016).

52 Migunov, V., Ryll, H., Zhuge, X., Simson, M., Struder, L., Batenburg, K.J., Houben, L. and Dunin-Borkowski, R.E. Rapid low dose electron tomography using a direct electron detection camera. *Scientific Reports* **5**, 14516 (2015).

53 Brilot, A.F., Chen, J.Z., Cheng, A., Pan, J., Harrison, S.C., Potter, C.S., Carragher, B., Henderson, R. and Grigorieff, N. Beam-induced motion of vitrified specimen on holey carbon film. *Journal of Structural Biology* **177**, 630–637 (2012).

54 Zheng, S.Q., Keszthelyi, B., Branlund, E., Lyle, J.M., Braunfeld, M.B., Sedat, J.W. and Agard, D.A. UCSF tomography: an integrated software suite for real-time electron

microscopic tomographic data collection, alignment, and reconstruction. *Journal of Structural Biology* **157**, 138–147 (2007).

55 Bharat, T.A.M., Hoffmann, P.C. and Kukulski, W. Correlative Microscopy of Vitreous Sections Provides Insights into BAR-Domain Organization In Situ. *Structure* (2018).

56 Celler, K., Koning, R.I., Willemse, J., Koster, A.J. and van Wezel, G.P. Cross-membranes orchestrate compartmentalization and morphogenesis in Streptomyces. *Nature Communications* 7, ncomms11836 (2016).

57 Tao, C.L., Liu, Y.T., Sun, R., Zhang, B., Qi, L., Shivakoti, S., Tian, C.L., Zhang, P., Lau, P.M., Zhou, Z.H. and Bi, G.Q. Differentiation and Characterization of Excitatory and Inhibitory Synapses by Cryo-electron Tomography and Correlative Microscopy. *Journal of Neuroscience* **38**, 1493–1510 (2018).

58 Wang, S., Li, S., Ji, G., Huang, X. and Sun, F. Using integrated correlative cryo-light and electron microscopy to directly observe syntaphilin-immobilized neuronal mitochondria in situ. *Biophysics Reports* **3**, 8–16 (2017).

59 Faas, F.G., Barcena, M., Agronskaia, A.V., Gerritsen, H.C., Moscicka, K.B., Diebolder, C.A., van Driel, L.F., Limpens, R.W., Bos, E., Ravelli, R.B., Koning, R.I. and Koster, A.J. Localization of fluorescently labeled structures in frozen-hydrated samples using integrated light electron microscopy. *Journal of Structural Biology* **181**, 283–290 (2013).

60 Koning, R.I., Faas, F.G., Boonekamp, M., de Visser, B., Janse, J., Wiegant, J.C., de Breij, A., Willemse, J., Nibbering, P.H., Tanke, H.J. and Koster, A.J. MAVIS: an integrated system for live microscopy and vitrification. *Ultramicroscopy* **143**, 67–76 (2014).

8

Cryo-Correlative Light and Electron Microscopy: Toward *in situ* Structural Biology

Tanmay A.M. Bharat[1] and Wanda Kukulski[2]

[1] Sir William Dunn School of Pathology, University of Oxford, United Kingdom
[2] Cell Biology Division, MRC Laboratory of Molecular Biology, Francis Crick Avenue, Cambridge, United Kingdom

8.1   Introduction

Electron microscopy (EM) is a unique method for imaging biological specimens, as it allows direct visualization of macromolecules in their native state (McDowall et al. 1983). Not only is the low-resolution envelope or shape of the macromolecule visible, but also near-atomic level information is preserved within EM images. This high-resolution information can be extracted from EM images by performing alignment and averaging, thereby allowing macromolecular structure determination (De Rosier and Klug, 1968). EM is equally compatible with cell biological investigations, providing high-resolution snapshots of cells and cellular components (Baumeister, 2002; Gan and Jensen, 2012). Despite numerous advantages, one of the major challenges in EM is locating macromolecules of interest in crowded and complex environments. The molecule of interest can be buried deep within the large volume of a cell or obscured by other components in large, multiprotein assemblies. One of the most popular methods to surmount this problem of locating molecules on EM grids is to combine the power of fluorescence microscopy (FM) with EM to locate specific macromolecules for selective and targeted imaging.

The current status of correlative light and electron microscopy (CLEM) methods has been a joint effort of both structural and cell biological electron microscopists. These two historically separate fields have therefore converged, and have found mutual interests. One result of this development is the recognition that there is a need to explore the complexity of macromolecular structures, assemblies of proteins, lipids and nucleic acids, in their native cellular environment (Beck and Baumeister 2016). This extends to visualizing the machinery necessary for functional and structural transitions, trajectories of conformational changes, rare and transient interactions, the supramolecular architecture of large assemblies, and the cellular microenvironment of biological processes. Obtaining these data requires imaging at various levels of complexity, which then integrate into a spatial and mechanistic understanding of cellular processes.

To study macromolecular structures in their native environment, it is essential to perform CLEM on frozen-hydrated (cryo) material, obtained through vitrification of the

Correlative Imaging: Focusing on the Future, First Edition. Edited by Paul Verkade and Lucy Collinson.
© 2020 John Wiley & Sons Ltd. Published 2020 by John Wiley & Sons Ltd.

sample (Lepault et al. 1983). As vitrification preserves cellular structures and molecular details to the atomic level, cryo-electron microscopy (cryo-EM) and cryo-electron tomography (cryo-ET) offer an opportunity for *in situ* structural biology where high-resolution structures of macromolecules can be resolved in cells (Briggs 2013). Single particle cryo-EM has recently been revolutionizing structural biology (Kuhlbrandt 2014), and the achievements of the cryo-EM field have been recognized with the 2017 Nobel Prize in Chemistry. These advances are now swapping over into cryo-ET, opening exciting perspectives when tied to developments in cryo-CLEM.

Figure 8.1 illustrates how different levels of biological complexity can be studied using cryo-EM and cryo-ET methods, from purified macromolecules to multicellular specimen such as tissues. We use filovirus assembly as a paradigm for illustrating the need for developing cryo-CLEM approaches at all levels, from *in vitro* to *in vivo*. Images represent cryo-CLEM applications at each conceptual level of biological complexity. In the first line of the chart under the images, we phrase specific, hypothetical questions regarding filovirus assembly that could potentially be answered using cryo-CLEM. In the last line, we outline the major areas of required development for this cryo-CLEM application to become viable.

In this chapter, we will discuss the potential of current and future cryo-CLEM approaches by systematically considering specimens with different levels of biological complexity, from purified samples in solution to macromolecular structures within multicellular tissues. At each level of biological complexity, we will describe the major issues that need to be overcome to achieve routine cryo-CLEM application. In most cases presented in this chapter, we make the implicit assumption that the goal of the experiment is to perform sub-tomogram averaging or single particle analysis for high-resolution structure determination after identifying the target feature on the EM grid using cryo-CLEM techniques.

8.2 Cryo-CLEM to Support Single Particle Analysis of Purified Macromolecules

In recent years, many structures from purified macromolecular complexes have been determined using single-particle cryo-EM methods (Fernandez-Leiro and Scheres 2016). Despite available routine procedures, challenges still arise, such as when very rare or transient complexes need to be identified within the sample. Cryo-CLEM has great potential to contribute to the study of such specimens by helping to identify complexes in different constitutional states. For example, a macromolecular complex tagged with a fluorophore could be mixed with a cofactor with low binding affinity that is tagged with another fluorophore before being frozen onto EM grids. Targeted cryo-EM imaging in positions where the two signals co-localize would then allow data collection on the transient complex for three-dimensional (3D) structure determination. The potential advantage of using this approach is obvious when the transient complex is present as a tiny fraction of the grid content (<0.1%), and therefore cannot be easily computationally isolated through classification procedures. We believe that even in cases where the complex is present in abundance (up to ~5%), this method will save valuable microscope and computing resources because imaging the entire grid and performing image reconstructions of non-cofactor bound complexes would be avoided.

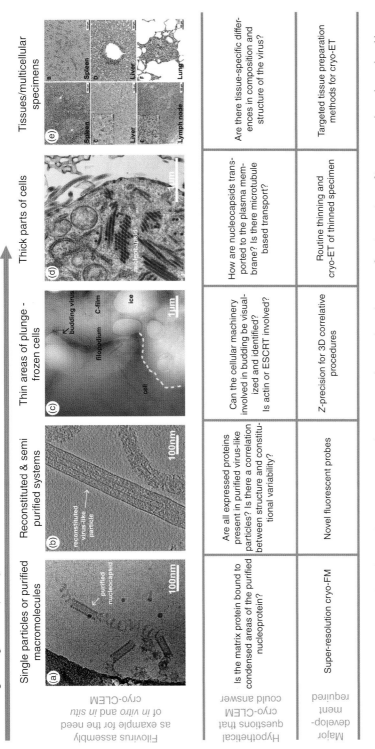

Figure 8.1 **(a)** Cryo-EM image of purified Ebola virus (EBOV) nucleoprotein and matrix (VP40) protein, forming regions of interspersed condensed and loose nucleocapsids. The areas of condensed nucleocapsids are predicted to contain nucleoprotein bound to matrix (Bharat et al. 2012). **(b)** Cryo-ET slice through an Ebola virus-like particle (VLP), reconstituted from nucleoprotein, matrix, and other accessory proteins (Bharat et al. 2012). It is not clear whether all VLPs contain the same ratio of components, a question appropriate for cryo-CLEM. **(c)** Cryo-EM of Marburg virus (MARV) budding from thin regions of infected cells. The cell outlines are indicated with yellow dashes. A virus (arrow) is budding from a filopodium. Image is modified from (Bharat et al. 2011), published under a CCBY licence: https://creativecommons.org/licenses/by/4.0/legalcode. **(d)** Room-temperature EM of resin-embedded cells infected with Ebola virus, showing assembled nucleocapsids in the cytoplasm. Image is modified from (Noda et al. 2006), published under a CCBY licence: https://creativecommons. org/licenses/by/4.0/legalcode. **(e)** Histology sections of tissue from different organs of Ebola virus-challenged animals. Upper left (spleen) and middle left (liver) images are untreated tissue; all others are anti-Ebola-virus immunoglobulin treated. Image is modified from (Dowall et al. 2017), published under a CCBY licence: https://creativecommons.org/licenses/by/4.0/legalcode.

The main challenge for cryo-CLEM of purified complexes is the fact that routine cryo-FM currently lacks both the sensitivity as well as the spatial resolution to identify single fluorophores and complexes. Although there are reports on localizing single fluorescent molecules under cryo-conditions (Li et al. 2015; Chang et al. 2014; Kaufmann et al. 2014), in a practical cryo-CLEM experiment, the current limits of detection in cryo-FM are probably in the range of dozens of fluorescent proteins. To address this challenge, tandem fluorophore repeats could be added to boost the signal from each tagged complex. However, this would increase the size of the fluorophore, possibly causing steric problems that might alter the characteristics of the complex being investigated, which would not be acceptable from a structural biology perspective. Thus, significant developments must be made to improve cryo-FM for this cryo-CLEM application to be viable.

Another, related challenge for cryo-CLEM application to single particles is that very high correlation precision, in the range of a few nanometres, will be required for this approach to work well, especially when cryo-EM imaging at high-magnification is needed. High magnification in cryo-EM means that the area of imaging would be small, making it crucial that the complex is localized within it, thus requiring high spatial precision. The magnification and pixel size in cryo-EM depend on the resolution targeted in single particle analysis, and therefore cannot be altered to accommodate the needs of cryo-CLEM. This can be illustrated by the following estimate: Since resolutions obtained in cryo-EM single-particle analysis are now better than 2 Å (Merk et al. 2016), many high-resolution data collections are conducted at a pixel size sampling of 0.5 Å at the specimen level. At this pixel size, the field of view on a standard 4K×4K detector is approximately 200 nm. Thus, the precision of localization must be better than that over the entire grid to ensure that the transient complexes of interest are not missed in the data collection, a big ask given the current state of cryo-CLEM.

Consequently, for cryo-CLEM to become useful for single-particle analysis, it will require advances in cryo super-resolution FM (Wolff et al. 2016). Such developments can build on recent efforts towards integrating super-resolution FM, cryo-FM and cryo-EM single-particle data collection (Kaufmann et al. 2014; Schorb and Briggs, 2014; Schorb et al. 2017). These recent developments are proof of principle studies that introduce hardware, software and sample handling, achieving improvements in localization precision, sample stability and ease of use.

One important challenge on the way to enable the exciting application of cryo-CLEM to single particles is the development of better objectives for cryo-FM that allow a small working distance and high numerical aperture, thus increasing resolution. To achieve this, the objective could for instance be placed in an immersion medium such as liquid propane or iso-pentane (Le Gros et al. 2009). Another requirement for obtaining higher spatial resolution in cryo-FM will be continued development and implementation of cryo-stages with high stability and negligible drift over the entirety of the imaging session (Wolff et al. 2016; Li et al. 2015; Chang et al. 2014).

An additional challenge for combining cryo-CLEM with structural analysis of single particles is the requirement of conducting cryo-FM imaging with minimal radiation damage, in order to maintain structural integrity to the atomic level. Radiation damage may be introduced by illumination with fluorescence light sources, in particular lasers, and would preclude high-resolution cryo-EM structure determination. Thus, cryo-FM may need to be carried out at low illumination power (Liu et al. 2015). The relationship between achievable resolution in cryo-EM and used laser intensities remains to be explored systematically. Thus, cryo-CLEM of single particles will also be illuminating from an academic aspect, and help us to understand how radiation damage by light affects the high-resolution structure of macromolecules.

8.3 Capturing Structural Dynamics of *in vitro* Reconstituted Systems

Biochemically reconstituted, dynamic systems lie on the border between cryo-EM and cryo-ET and they present slightly different challenges for cryo-CLEM studies compared to purified single complexes. The types of specimens we refer to are protein, membrane, and nucleic acid assemblies, including reconstituted coated vesicles (Dodonova et al. 2015), cytoskeletal assemblies (Sandblad et al. 2006), and chromatin (Scheffer et al. 2011). We particularly point to reconstituted systems where a dynamic process is being studied, such as SNARE-mediated membrane fusion (Bharat et al. 2014) or clathrin-mediated membrane budding (Dannhauser and Ungewickell 2012). These dynamic processes often involve key structural rearrangements that are mechanistically important and need high-resolution structural information for detailed understanding of the underlying biological process. In such cases, the use of cryo-CLEM supplies critical information about the time point or state in which the structure is being observed.

Cryo-FM and cryo-EM are both comparatively easy to apply to these systems, with the usual caveats of sample preparation and optimization. Specimen thickness is generally maximally up to a few hundreds of nanometres and the structures of interest are often easy to identify, such as a coated vesicle (Dodonova et al. 2015). The magnification and thus the field of view are generally less constrained than for single particles. These properties of the samples are advantageous for cryo-FM, and therefore in many cases of reconstituted or semi-purified cellular components, cryo-CLEM should be readily applicable without requiring fundamental developments.

However, challenges arise when a particular event is being researched, such as the key moment of membrane fusion or vesicle budding, while other states of the system are not as interesting, or are not the main goal of the inquiry. Furthermore, the different states (i.e., the various stages of a fusion process) may not be easily discernible visually based on composition or overall structure. The difficulty in these cases is to find an appropriate biochemical probe that shows a characteristic fluorescent signature only when the key event of interest occurs during the dynamic biological process.

Thus cryo-CLEM of *in vitro* reconstituted or semipurified dynamic systems will benefit greatly from the development of novel fluorescent probes that define precise time points in biological processes, and will require major time investments into optimizing the biochemistry of the sample. Once such systems are established, they will provide great opportunities to structurally access transient and dynamic steps during reconstituted cellular processes.

There are several examples of existing labels that could fulfil this future purpose. For example, a possible way to detect when two components come together to carry out a biochemical process in a dynamic reaction could be to use labels such as split-GFP (Cabantous et al. 2005). In this case, the detection of a fluorescent signal in cryo-FM could be directly linked to the presence of a transient protein–protein interaction that had been formed at the time point of freezing, allowing the experimentalist to access a desired state of the system. Other ways to obtain fluorescent signatures could be to use click chemistry-fluorescent probes that become active when a reaction takes place, such as lipid mixing in a membrane fusion assay with differently functionalized lipids in apposing membranes (Malsam et al. 2012).

Another future application might be to study pulse-chase labeled reactions that could be imaged after a defined time point extrapolated from live imaging. This option could be explored with a HaloTag, for example (Los et al. 2008). Fluorescent ligands to the HaloTag could be added to the specimen and plunge-frozen at set time points, when the tag has become active inside the *in vitro* reconstituted system, followed by cryo-CLEM. This kind of tagging approach will be limited by the ability of the fluorescent ligand to access the tagged site during the dynamic reaction being investigated, however, changes in accessibility could even be exploited to study the timing of structural rearrangements. Along the same lines, the use of caged components (Ellis-Davies, 2007) for *in situ* uncaging reactions by a light pulse very shortly before plunge freezing could be used to capture or enrich fast reactions on the EM grid. A similar approach has been previously used for cryo-EM structure determination of complexes in an activated state by spraying a chemical onto the EM grid shortly before plunge freezing (Berriman and Unwin, 1994) and further in an integrated FM and plunging device (Koning et al. 2014b). After the plunge-freezing step, cryo-FM could be used to identify those events where the process has occurred, allowing target selection for cryo-EM. The development and implementation of novel chemical probes is outside the normal expertise of most CLEM experts, thus strong collaborations must be established with chemical biologists and chemists for such workflows to be developed.

8.4 Identifying Macromolecules in Plunge-Frozen Whole Cells

Many recent cryo-CLEM applications have focused on plunge-frozen cell samples, when either thin bacterial cells (Koning et al. 2014a) or thin areas of eukaryotic cells were under investigation (Schellenberger et al. 2014). Imaging plunge-frozen whole cells is the traditional wheelhouse of cryo-CLEM and protocols have arguably been best developed for applications at this level of biological complexity (Hampton et al. 2017). That being said, cryo-CLEM of plunge-frozen cells could be further improved in the future, and will benefit most from developments for better correlation precision in 3D as well as image processing approaches for macromolecular detection and structure determination in crowded cellular environments.

The types of specimens in this category are typically less than a micron thick, and include bacterial cells, thin areas of eukaryotic cells such as filopodia and cortical regions of cells. This category may also include crude cell extracts containing large organelles. Due to significant specimen volume, structures overlap in 2D projection images, and therefore these samples are generally imaged in 3D by cryo-ET. There are good reasons for visualizing macromolecular complexes within their cellular environment by cryo-ET, rather than in isolation: Some structures, or interactions between complexes and cofactors, are unique, very transient and unstable and therefore may only exist in the context of a cell or an intact organelle, such as the exocyst complex

(Picco et al. 2017), the arrangement of ATP synthases in mitochondrial membranes (Davies et al. 2012), or the bacterial divisome (Szwedziak et al. 2014).

A first goal for cryo-CLEM in this context is to localize macromolecular structures with high accuracy. The challenge is set by the 3D crowding within the cellular volume, requiring highly precise correlation between cryo-FM and cryo-EM in the z-dimension. For this aim to be achieved, image processing and landmarking approaches need to be developed for performing 3D registrations between cryo-FM and cryo-EM data with an isotropic accuracy of approximately 50 nm. Cryo-confocal FM provides focal images of ~300 nm in depth (Arnold et al. 2016), therefore major efforts need to be made to achieve improved z-localization. Like single particle cryo-CLEM, this application will thus benefit from developments in cryo-superresolution imaging (Chang et al. 2014). It may also require careful deposition of fiducials to the cellular specimen so that they are evenly distributed into the entirety of the volume, perhaps not only on the cell but also inside the cell. Furthermore, software supporting on-the-fly correlation (Paul-Gilloteaux et al. 2017) and image processing will increase the throughput of routine applications. While discussing these future directions, it is pertinent to mention the efforts from many labs that have paved the way for the field, enabling many cell biological questions to be already tackled by existing cryo-CLEM methodology using plunge-frozen whole cells (van Driel et al. 2009, Schwartz et al. 2007, Briegel et al. 2010, Sartori et al. 2007).

Once the location of the macromolecule or complex has been narrowed down using cryo-CLEM to a cubic subvolume of approximately 50 nm side length, a related image processing challenge will be to unambiguously identify it in the crowded and complicated EM density volume, and to determine its orientation. This has so far been achieved using template-matching methods (Frangakis et al. 2002; Rickgauer et al. 2017), where a high-resolution structure is compared with boxes extracted from the tomogram yielding a cross-correlation map, which could be read as probabilities for finding the target in the tomogram. While this approach will hopefully be further improved and combined with cryo-CLEM, it is currently mostly applicable to complexes of known or predictable structure with very characteristic overall shapes, such as proteasomes or ribosomes (Asano et al. 2015; Pfeffer et al. 2015). Our vision for identification of macromolecules of elusive and unknown structure, such as very transiently interacting complexes, would be a procedure where the signal obtained by cryo-FM would be used to direct and drive macromolecular identification inside cells without the need for a template. Based on the cryo-FM signal, subvolumes could then be extracted from tomograms and then subjected to pattern-recognition approaches to identify repeating shapes, allowing identification of as-yet-unknown structures.

Finally, after successful identification, the complexes will need to be averaged using sub-tomogram averaging approaches for high-resolution structure determination and biochemical interpretation from fitted atomic models. While recent reports show the power of sub-tomogram averaging inside crowded cells (Bykov et al. 2017), there is still a long way to go until atomic level resolution will be routinely achievable because the signal-to-noise ratio in cellular tomograms is inherently low (Asano et al. 2016). Possible ways to apply sub-tomogram averaging in these cases could be through the use of novel refinement approaches such as regularized-likelihood optimization or other constrained cross-correlation approaches (Förster et al. 2005; Bartesaghi and Subramaniam, 2009; Bharat and Scheres, 2016). Another method to improve contrast in cellular tomograms is the use of phase-plates in high-end microscopes allowing imaging of the specimen without defocus (Fukuda et al. 2015).

8.5 Macromolecular Structures in Thinned Samples from Thick Cell Areas

Only a limited number of cellular processes occur in cellular regions that are thin enough to be directly imaged by cryo-ET. Therefore, cell areas in the range of 1 μm thickness or more need to be first thinned either using cryo-ultramicrotomy (aka cryo EM of vitreous sections; CEMOVIS) or focused ion beam (FIB) milling in a dual-beam FIB-Scanning EM to produce so-called lamella (Dubochet et al. 1988; Marko et al. 2007). These methods allow, in otherwise inaccessible parts of the cell, imaging the native organization of protein and nucleic acid assemblies that cannot be isolated or reconstituted without loss of integrity, such as large organelle membranes or the inside of the nucleus (Mahamid et al. 2016; Scheffer et al. 2011). Because the achievable resolution in cryo-ET directly depends on thickness, thinning may be desirable even for samples where imaging would in principle also be possible without thinning, for example when protein bundles or arrays in bacteria need to be imaged at high resolution (Bharat et al. 2017; Bharat et al. 2015; Salje et al. 2009). The reasons for implementing cryo-CLEM for these kinds of projects are similar to those for thin cellular samples: Localization and identification of elusive protein assemblies to perform their structural analysis *in situ*.

Cryo-FM has already been successfully used for targeted thinning by FIB-milling of cell regions that contain fluorescent signals of interest (Arnold et al. 2016). This approach ensures that the cell to be thinned and subsequently imaged has been, for instance, successfully transfected. For precise localization of small or elusive protein assemblies and streamlined structural analysis thereafter, additional developments are needed. Current bottlenecks relate to the fact that both vitreous sectioning and FIB milling are technically demanding and time-consuming. Obtaining a thin sample of a thick cellular region, which is structurally preserved to the atomic level, is further complicated by the fact that both methods are prone to artefacts.

Vitreous sections suffer from cutting (knife)-induced compression damage and crevasses on the section surface, deteriorating the structural integrity and obscuring the field of view, respectively (Al-Amoudi et al. 2005). While some suggestions have been made in order to reduce or correct for these artefacts (Al-Amoudi et al. 2003; Pierson et al. 2011), the remaining effects on high-resolution information cannot be ignored. In contrast to vitreous sectioning, which uses high-pressure frozen starting material, FIB-milling relies on plunge-frozen cells adhered to EM grids, meaning that vitrification is limited to a couple of microns in thickness, and very thick areas of cells such as large nuclei may not be vitrified (Iancu et al. 2006). A further inconvenience is that vitrification can only be assessed after the tedious FIB milling process, in the transmission EM. Thus, it would be desirable to have improved vitrification procedures for thick cellular samples on EM grids, such as with an optimized plunge-freezing device. Furthermore, the lamella produced by FIB-milling may suffer from beam damage induced by the gallium ion beam, but this has not yet been systematically explored (Marko et al. 2007).

Depending on the vitrification and thinning method, cryo-FM could be performed before or after sample thinning. Vitreous sections have been successfully imaged by cryo-FM, and fluorescent fiducial markers could be adhered to the grid for high-precision correlation (Gruska et al. 2008; Schorb et al. 2017). More recently, a workflow has been proposed to generate vitreous sections from plunge-frozen cells on EM grids, which could potentially be combined with cryo-FM prior to vitreous sectioning (Kolovou et al. 2017). For rare structures, it is helpful to estimate the occurrence of the signal of interest per grid: Side lengths of vitreous sections are in the range of 50–100 μm. Dozens of grid squares can be covered with sections, such that the material on one grid would sample roughly 10 μm of equivalent specimen depth. This means that the specimen volume screened by cryo-FM of one grid is in principle large enough to screen 5–10 mammalian tissue culture cells, or hundreds of yeast cells.

In the case of FIB-milling, the small field of view opened up by the lamella can be a major limitation: Lamella are typically about 15–20 μm in side length, comprising usually not more than one mammalian cell or 5–8 yeast cell profiles, and rarely more than 3 lamella are made on one grid (Villa et al. 2013). This means that sparsely distributed structures will be rare to find. It is possible to perform FIB-milling guided by 3D cryo-CLEM, using fluorescent fiducial markers and cryo-FM images acquired prior to transfer into the FIB-SEM (Arnold et al. 2016). However, the accuracy to determine lamella position in the z-dimension is in the micron range, thus structures smaller than the lamella thickness of ~200 nm will be frequently missed. Another possibility, not yet implemented, would be integrated fluorescence imaging during FIB-milling as guidance for targeted milling. Cryo-FM of lamella post-milling has also so far not been reported, but would allow the confirmation of presence as well as localization of the signal, especially for ubiquitous signals that do not require targeted milling.

Improvements at the level of cryo-ET data acquisition are also required. Vitreous sections can be poorly attached to the carbon support and therefore display erratic movement upon exposure to the electron beam, making data acquisition nearly impossible (Sartori Blanc et al. 1998; Pierson et al. 2010). Evenness and attachment of vitreous sections could potentially be assessed by cryo-FM. Lamella can display beam-induced movement due to charging, which could be addressed by deposition of heavy metals such as platinum, shown to reduce charging during volta phase plate imaging (Mahamid et al. 2016). Finally, the achievable resolution depends on the alignment of the tilt series, which is ideally performed using high-contrast fiducial markers such as gold beads (Amat et al. 2010). Procedures to deposit fiducial markers on vitreous sections and lamella have been proposed (Gruska et al. 2008; Masich et al. 2006; Harapin et al. 2015), and could serve to develop routine protocols. Currently, in most cases, aligning tilt series from vitreous sections and lamella relies on fiducial-less algorithms (Castano-Diez et al. 2010; Mastronarde and Held, 2017; Noble and Stagg, 2015; Sorzano et al. 2009). How compatible these alignment methods are with the requirements for subnanometer-resolution structural determination remains to be assessed, and possibly improved.

8.6 Enabling Structural Biology in Multicellular Organisms and Tissues by Cryo-CLEM

Determining macromolecular structures within specialized cells of native tissues would open up new avenues for addressing tissue-dependent structural and compositional variability of macromolecular assemblies, or differences in supramolecular organization. Such studies would be a major step toward elucidating the role of macromolecules in tissue development and physiology, in health and in disease. Possible target samples would include whole multicellular organisms or primary organ tissue derived from biopsy samples from animal models or patients. The potential of structural biology within tissue has been demonstrated in a pioneering study that determined the structure of cadherin in human skin desmosomes through sub-tomogram averaging, using cryo-ET data of vitreous sections (Al-Amoudi et al. 2007).

There are two major reasons for employing cryo-CLEM on multicellular samples. The first is to locate the cell of interest containing the fluorescent signal in a sometimes bewilderingly complicated and large 3D tissue. Thus, when the macromolecule of interest is expressed only in a subset of cells in the tissue, the target region or cell could be identified using fluorescent signals. The second reason is that once the cell (or cells) of interest has been located, cryo-CLEM would permit subcellular localization of macromolecular complexes within the cells.

While these two reasons may appear to be addressable in a single cryo-CLEM step, as described later, they will likely form two separate steps in a future cryo-CLEM procedure. Cryo-CLEM of tissues is currently far from routine. Its implementation is prevented by a number of challenges, and any future workflow will probably involve a complicated multistep procedure. Difficulties arise mostly from the sheer size and complexity of the starting material. First, only high-pressure freezing can vitrify multicellular specimen up to 300 µm in thickness (Dubochet, 1995). Protocols yielding consistently vitrified samples are available for a number of model organisms and tissues (McDonald et al. 2010), but not for all. It can be difficult to produce a 300 µm thin sample, such as a biopsy punch, and transfer it into the HPF without affecting structural integrity. For instance, in the case of fresh brain tissue, it has been shown using freeze substitution that very rapid preparation for high-pressure freezing is key to preserve ultrastructure (Korogod et al. 2015). As the cryo-CLEM experiments we have in mind aim at macromolecular structure determination, it is critical that structural integrity is maintained to the atomic level. Possible protocols may include vibratome sectioning prior to high-pressure freezing (Sartori Blanc et al. 1998).

Thinning tissue samples for cryo-EM is what vitreous sectioning traditionally has been used for. Several representative studies show that if a tissue can be high-pressure frozen, it will generally also be amenable to cryo-EM of vitreous sections (Zuber et al. 2005; Hsieh et al. 2006; Fernandez-Busnadiego et al. 2010; Gunkel et al. 2015). FIB-milling, on the other hand, is optimized for plunge-frozen single cells adhered to EM grids, but two possible approaches have been proposed for high-pressure frozen, multicellular samples. The so-called lift-out approach consists of two FIB-milling steps, first cutting out a lamella of few micrometers in thickness from the bulk specimen, placing it onto the EM grid followed by further thinning (Mahamid et al. 2015). Another approach suggests using a single step to FIB-mill organisms that were high-pressure frozen on EM grids (Harapin et al. 2015). While both proof-of-principle studies were performed using the relatively small *C. elegans*, the lift-out approach, if the workflow becomes routine, is more likely to be applicable to bulkier specimen, as FIB-milling times would become impractically long for removal of large chunks of specimen. Cryo-planing and pre-trimming are other options to roughly pre-thin samples prior to FIB-milling; these approaches suggest using cryo-microtomy to remove bulk material from the EM grid or from high-pressure frozen samples (Rigort et al. 2010; Hsieh et al. 2014).

A first cryo-CLEM step could help to localize the cell or area of interest within the full tissue volume, requiring precision of ~1 µm. This step would guide the thinning procedure to the target position. Because obtaining vitreous sections or lamella is tedious and time consuming, it is crucial to direct the thinning to the correct region within the large sample generated by high-pressure freezing. The lift-out approach described above has indeed been using cryo-FM images for guidance during the rough lamella generation step (Mahamid et al. 2015). Depending on availability of visual cues, and the dimensions of the part of the tissue that will contain the structures of interest, cryo-FM images will need to be 3D focal stacks, and therefore this targeting step will also benefit from improvements in 3D correlation precision. There are 3D CLEM solutions for resin-embedded tissues (Kolotuev et al. 2009; Karreman et al. 2014), which may serve as starting points to develop procedures for cryo samples.

The second cryo-FM step would be used after sample thinning to subcellularly localize the structure of interest with high precision, in a similar way as described for thick single cells. There may be potential for combining the two cryo-CLEM steps, for instance if it were possible to develop an integrated sectioning and fluorescence microscopy device at cryo-temperature that ideally could produce section thicknesses of less than 500 nm (Li et al. 2010).

8.7 Conclusions

There is a growing demand among structural and cell biologists to study the structures of macromolecular assemblies formed by protein, lipid and nucleic acid complexes, in increasingly native situations. This requires routine imaging of structures as purified components in solution, in reconstituted systems, in single cell culture models and in multicellular samples such as tissues and organisms. The degree of nativeness goes hand in hand with the level of complexity, and with different needs for localizing, identifying or discriminating the structures of interest. Information on the dynamics of structural rearrangements and on local cellular composition is also fundamental in order to understand how macromolecular assemblies function. Cryo-CLEM is therefore ideally poised to implement workflows for *in situ* structural biology. Developments are needed at many different levels; improvements in sample preparation are as important as advances in cryo-FM imaging, in reliable correlation procedures, in improved cryo-ET acquisition and in novel software implementations for image processing. These developments promise a bright future for the cryo-CLEM field.

Acknowledgments

We would like to thank Abul Tarafder and Errin Johnson for helpful comments on the manuscript. TAMB is a recipient of a Sir Henry Dale Fellowship jointly funded by the Wellcome Trust and the Royal Society (Grant Number 202231/Z/16/Z). Work in the group of WK is supported by the Medical Research Council (MC_UP_1201/8).

References

Al-Amoudi, A., Diez, D. C., Betts, M. J., and Frangakis, A. S. 2007. The molecular architecture of cadherins in native epidermal desmosomes. *Nature*, **450**, 832–837.

Al-Amoudi, A., Dubochet, J., Gnaegi, H., Luthi, W., and Studer, D. 2003. An oscillating cryo-knife reduces cutting-induced deformation of vitreous ultrathin sections. *J Microsc*, **212**, 26–33.

Al-Amoudi, A., Studer, D., and Dubochet, J. 2005. Cutting artefacts and cutting process in vitreous sections for cryo-electron microscopy. *J Struct Biol*, **150**, 109–121.

Amat, F., Castaño-Diez, D., Lawrence, A., Moussavi, F., Winkler, H., and Horowitz, M. 2010. Alignment of cryo-electron tomography datasets. *Methods Enzymol*, **482**, 343–367.

Arnold, J., Mahamid, J., Lucic, V., De Marco, A., Fernandez, J. J., Laugks, T., Mayer, T., Hyman, A. A., Baumeister, W., and Plitzko, J. M. 2016. Site-Specific Cryo-focused Ion Beam Sample Preparation Guided by 3D Correlative Microscopy. *Biophys J*, **110**, 860–869.

Asano, S., Engel, B. D., and Baumeister, W. 2016. In Situ Cryo-Electron Tomography: A Post-Reductionist Approach to Structural Biology. *J Mol Biol*, **428**, 332–343.

Asano, S., Fukuda, Y., Beck, F., Aufderheide, A., Forster, F., Danev, R., and Baumeister, W. 2015. Proteasomes. A molecular census of 26S proteasomes in intact neurons. *Science*, **347**, 439–442.

Bartesaghi, A., and Subramaniam, S. 2009. Membrane protein structure determination using cryo-electron tomography and 3D image averaging. *Curr Opin Struct Biol*, **19**, 402–407.

Baumeister, W. 2002. Electron tomography: towards visualizing the molecular organization of the cytoplasm. *Curr Opin Struct Biol*, **12**, 679–684.

Beck, M., and Baumeister, W. 2016. Cryo-Electron Tomography: Can it Reveal the Molecular Sociology of Cells in Atomic Detail? *Trends Cell Biol*, **26**, 825–837.

Berriman, J., and Unwin, N. 1994. Analysis of transient structures by cryo-microscopy combined with rapid mixing of spray droplets. *Ultramicroscopy*, **56**, 241–252.

Bharat, T. A., Malsam, J., Hagen, W. J., Scheutzow, A., Sollner, T. H., and Briggs, J. A. 2014. SNARE and regulatory proteins induce local membrane protrusions to prime docked vesicles for fast calcium-triggered fusion. *EMBO Rep*, **15**, 308–314.

Bharat, T. A., Murshudov, G. N., Sachse, C., and Lowe, J. 2015. Structures of actin-like ParM filaments show architecture of plasmid-segregating spindles. *Nature*, **523**, 106–110.

Bharat, T. A., Noda, T., Riches, J. D., Kraehling, V., Kolesnikova, L., Becker, S., Kawaoka, Y., and Briggs, J. A. 2012. Structural dissection of Ebola virus and its assembly determinants using cryo-electron tomography. *Proc Natl Acad Sci U S A*, **109**, 4275–4280.

Bharat, T. A., Riches, J. D., Kolesnikova, L., Welsch, S., Krahling, V., Davey, N., Parsy, M. L., Becker, S., and Briggs, J. A. 2011. Cryo-electron tomography of Marburg virus particles and their morphogenesis within infected cells. *PLoS Biol*, **9**, e1001196.

Bharat, T. A. M., Kureisaite-Ciziene, D., Hardy, G. G., Yu, E. W., Devant, J. M., Hagen, W. J. H., Brun, Y. V., Briggs, J. A. G., and Lowe, J. 2017. Structure of the hexagonal surface layer on Caulobacter crescentus cells. *Nat Microbiol*, **2**, 17059.

Bharat, T. A. M., and Scheres, S. H. W. 2016. Resolving macromolecular structures from electron cryo-tomography data using subtomogram averaging in RELION. *Nature Protocols*, **11**, 9–20.

Briegel, A., Chen, S., Koster, A. J., Plitzko, J. M., Schwartz, C. L., and Jensen, G. J. 2010. Correlated light and electron cryo-microscopy. *Methods Enzymol*, **481**, 317–341.

Briggs, J. A. 2013. Structural biology in situ – the potential of subtomogram averaging. *Curr Opin Struct Biol*, **23**, 261–267.

Bykov, Y. S., Schaffer, M., Dodonova, S. O., Albert, S., Plitzko, J. M., Baumeister, W., Engel, B. D., and Briggs, J. A. 2017. The structure of the COPI coat determined within the cell. *Elife*, **6**.

Cabantous, S., Terwilliger, T. C., and Waldo, G. S. 2005. Protein tagging and detection with engineered self-assembling fragments of green fluorescent protein. *Nat Biotechnol*, **23**, 102–107.

Castano-Diez, D., Scheffer, M., Al-Amoudi, A., and Frangakis, A. S. 2010. Alignator: a GPU powered software package for robust fiducial-less alignment of cryo tilt-series. *J Struct Biol*, **170**, 117–126.

Chang, Y. W., Chen, S., Tocheva, E. I., Treuner-Lange, A., Lobach, S., Sogaard-Andersen, L., and Jensen, G. J. 2014. Correlated cryogenic photoactivated localization microscopy and cryo-electron tomography. *Nat Methods*, **11**, 737–739.

Dannhauser, P. N., and Ungewickell, E. J. 2012. Reconstitution of clathrin-coated bud and vesicle formation with minimal components. *Nat Cell Biol*, **14**, 634–639.

Davies, K. M., Anselmi, C., Wittig, I., Faraldo-Gomez, J. D., and Kuhlbrandt, W. 2012. Structure of the yeast F1Fo-ATP synthase dimer and its role in shaping the mitochondrial cristae. *Proc Natl Acad Sci U S A*, **109**, 13602–13607.

De Rosier, D. J., and Klug, A. 1968. Reconstruction of three dimensional structures from electron micrographs. *Nature*, **217**, 130–134.

Dodonova, S. O., Diestelkoetter-Bachert, P., Von Appen, A., Hagen, W. J., Beck, R., Beck, M., Wieland, F., and Briggs, J. A. 2015. VESICULAR TRANSPORT. A structure of the COPI coat and the role of coat proteins in membrane vesicle assembly. *Science*, **349**, 195–198.

Dowall, S. D., Jacquot, F., Landon, J., Rayner, E., Hall, G., Carbonnelle, C., Raoul, H., Pannetier, D., Cameron, I., Coxon, R., Al Abdulla, I., Hewson, R., and Carroll, M. W. 2017. Post-exposure treatment of non-human primates lethally infected with Ebola virus with EBOTAb, a purified ovine IgG product. *Sci Rep*, 7, 4099.

Dubochet, J. 1995. High-pressure freezing for cryoelectron microscopy. *Trends Cell Biol*, **5**, 366–368.

Dubochet, J., Adrian, M., Chang, J. J., Homo, J. C., Lepault, J., Mcdowall, A. W., and Schultz, P. 1988. Cryo-electron microscopy of vitrified specimens. *Q Rev Biophys*, **21**, 129–228.

Ellis-Davies, G. C. 2007. Caged compounds: photorelease technology for control of cellular chemistry and physiology. *Nat Methods*, **4**, 619–628.

Fernandez-Busnadiego, R., Zuber, B., Maurer, U. E., Cyrklaff, M., Baumeister, W., and Lucic, V. 2010. Quantitative analysis of the native presynaptic cytomatrix by cryoelectron tomography. *J Cell Biol*, **188**, 145–156.

Fernandez-Leiro, R., and Scheres, S. H. 2016. Unravelling biological macromolecules with cryo-electron microscopy. *Nature*, **537**, 339–346.

Förster, F., Medalia, O., Zauberman, N., Baumeister, W., and Fass, D. 2005. Retrovirus envelope protein complex structure in situ studied by cryo-electron tomography. *Proc Natl Acad Sci U S A*, **102**, 4729–4734.

Frangakis, A. S., Bohm, J., Forster, F., Nickell, S., Nicastro, D., Typke, D., Hegerl, R., and Baumeister, W. 2002. Identification of macromolecular complexes in cryoelectron tomograms of phantom cells. *Proc Natl Acad Sci U S A*, **99**, 14153–14158.

Fukuda, Y., Laugks, U., Lucic, V., Baumeister, W., and Danev, R. 2015. Electron cryotomography of vitrified cells with a Volta phase plate. *J Struct Biol*, **190**, 143–154.

Gan, L., and Jensen, G. J. 2012. Electron tomography of cells. *Q Rev Biophys*, **45**, 27–56.

Gruska, M., Medalia, O., Baumeister, W., and Leis, A. 2008. Electron tomography of vitreous sections from cultured mammalian cells. *J Struct Biol*, **161**, 384–392.

Gunkel, M., Schoneberg, J., Alkhaldi, W., Irsen, S., Noe, F., Kaupp, U. B., and Al-Amoudi, A. 2015. Higher-order architecture of rhodopsin in intact photoreceptors and its implication for phototransduction kinetics. *Structure*, **23**, 628–638.

Hampton, C. M., Strauss, J. D., Ke, Z., Dillard, R. S., Hammonds, J. E., Alonas, E., et al. 2017. Correlated fluorescence microscopy and cryo-electron tomography of virus-infected or transfected mammalian cells. *Nat Protoc*, **12**, 150–167.

Harapin, J., Bormel, M., Sapra, K. T., Brunner, D., Kaech, A., and Medalia, O. 2015. Structural analysis of multicellular organisms with cryo-electron tomography. *Nat Methods*, **12**, 634–636.

Hsieh, C., Schmelzer, T., Kishchenko, G., Wagenknecht, T., and Marko, M. 2014. Practical workflow for cryo focused-ion-beam milling of tissues and cells for cryo-TEM tomography. *J Struct Biol*, **185**, 32–41.

Hsieh, C. E., Leith, A., Mannella, C. A., Frank, J., and Marko, M. 2006. Towards high-resolution three-dimensional imaging of native mammalian tissue: electron tomography of frozen-hydrated rat liver sections. *J Struct Biol*, **153**, 1–13.

Iancu, C. V., Tivol, W. F., Schooler, J. B., Dias, D. P., Henderson, G. P., Murphy, G. E., Wright, E. R., Li, Z., Yu, Z., Briegel, A., Gan, L., He, Y., and Jensen, G. J. 2006. Electron cryotomography sample preparation using the Vitrobot. *Nat Protoc*, **1**, 2813–2819.

Karreman, M. A., Mercier, L., Schieber, N. L., Shibue, T., Schwab, Y., and Goetz, J. G. 2014. Correlating intravital multi-photon microscopy to 3D electron microscopy of invading tumor cells using anatomical reference points. *PLoS One*, **9**, e114448.

Kaufmann, R., Schellenberger, P., Seiradake, E., Dobbie, I. M., Jones, E. Y., Davis, I., Hagen, C., and Grunewald, K. 2014. Super-resolution microscopy using standard fluorescent proteins in intact cells under cryo-conditions. *Nano Lett*, **14**, 4171–4175.

Kolotuev, I., Schwab, Y., and Labouesse, M. 2009. A precise and rapid mapping protocol for correlative light and electron microscopy of small invertebrate organisms. *Biol Cell*, **102**, 121–132.

Kolovou, A., Schorb, M., Tarafder, A., Sachse, C., Schwab, Y., and Santarella-Mellwig, R. 2017. A new method for cryo-sectioning cell monolayers using a correlative workflow. *Methods Cell Biol*, **140**, 85–103.

Koning, R. I., Celler, K., Willemse, J., Bos, E., Van Wezel, G. P., and Koster, A. J. 2014a. Correlative cryo-fluorescence light microscopy and cryo-electron tomography of Streptomyces. *Methods Cell Biol*, **124**, 217–239.

Koning, R. I., Faas, F. G., Boonekamp, M., De Visser, B., Janse, J., Wiegant, J. C., De Breij, A., Willemse, J., Nibbering, P. H., Tanke, H. J., and Koster, A. J. 2014b. MAVIS: an integrated system for live microscopy and vitrification. *Ultramicroscopy*, **143**, 67–76.

Korogod, N., Petersen, C. C., and Knott, G. W. 2015. Ultrastructural analysis of adult mouse neocortex comparing aldehyde perfusion with cryo fixation. *Elife*, **4**.

Kuhlbrandt, W. 2014. Biochemistry. The resolution revolution. *Science*, **343**, 1443–1444.

Le Gros, M. A., Mcdermott, G., Uchida, M., Knoechel, C. G., and Larabell, C. A. 2009. High-aperture cryogenic light microscopy. *J Microsc*, **235**, 1–8.

Lepault, J., Booy, F. P., and Dubochet, J. 1983. Electron microscopy of frozen biological suspensions. *J Microsc*, **129**, 89–102.

Li, A., Gong, H., Zhang, B., Wang, Q., Yan, C., Wu, J., Liu, Q., Zeng, S., and Luo, Q. 2010. Micro-optical sectioning tomography to obtain a high-resolution atlas of the mouse brain. *Science*, **330**, 1404–1408.

Li, W., Stein, S. C., Gregor, I., and Enderlein, J. 2015. Ultra-stable and versatile widefield cryo-fluorescence microscope for single-molecule localization with sub-nanometer accuracy. *Opt Express*, **23**, 3770–3783.

Liu, B., Xue, Y., Zhao, W., Chen, Y., Fan, C., Gu, L., Zhang, Y., Zhang, X., Sun, L., Huang, X., Ding, W., Sun, F., Ji, W., and Xu, T. 2015. Three-dimensional super-resolution protein localization correlated with vitrified cellular context. *Sci Rep*, **5**, 13017.

Los, G. V., Encell, L. P., Mcdougall, M. G., Hartzell, D. D., Karassina, N., Zimprich, C., et al. 2008. HaloTag: a novel protein labeling technology for cell imaging and protein analysis. *ACS Chem Biol*, **3**, 373–382.

Mahamid, J., Pfeffer, S., Schaffer, M., Villa, E., Dancv, R., Cuellar, L. K., et al. 2016. Visualizing the molecular sociology at the HeLa cell nuclear periphery. *Science*, **351**, 969–972.

Mahamid, J., Schampers, R., Persoon, H., Hyman, A. A., Baumeister, W., and Plitzko, J. M. 2015. A focused ion beam milling and lift-out approach for site-specific preparation of frozen-hydrated lamellas from multicellular organisms. *J Struct Biol*, **192**, 262–269.

Malsam, J., Parisotto, D., Bharat, T. A., Scheutzow, A., Krause, J. M., Briggs, J. A., and Sollner, T. H. 2012. Complexin arrests a pool of docked vesicles for fast Ca2+-dependent release. *EMBO J*, **31**, 3270–3281.

Marko, M., Hsieh, C., Schalek, R., Frank, J., and Mannella, C. 2007. Focused-ion-beam thinning of frozen-hydrated biological specimens for cryo-electron microscopy. *Nat Methods*, **4**, 215–217.

Masich, S., Ostberg, T., Norlen, L., Shupliakov, O., and Daneholt, B. 2006. A procedure to deposit fiducial markers on vitreous cryo-sections for cellular tomography. *J Struct Biol*, **156**, 461–468.

Mastronarde, D. N., and Held, S. R. 2017. Automated tilt series alignment and tomographic reconstruction in IMOD. *J Struct Biol*, **197**, 102–113.

Mcdonald, K., Schwarz, H., Muller-Reichert, T., Webb, R., Buser, C., and Morphew, M. 2010. "Tips and tricks" for high-pressure freezing of model systems. *Methods Cell Biol*, **96**, 671–693.

Mcdowall, A. W., Chang, J. J., Freeman, R., Lepault, J., Walter, C. A., and Dubochet, J. 1983. Electron microscopy of frozen hydrated sections of vitreous ice and vitrified biological samples. *J Microsc*, **131**, 1–9.

Merk, A., Bartesaghi, A., Banerjee, S., Falconieri, V., Rao, P., Davis, M. I., Pragani, R., Boxer, M. B., Earl, L. A., Milne, J. L., and Subramaniam, S. 2016. Breaking Cryo-EM Resolution Barriers to Facilitate Drug Discovery. *Cell*, **165**, 1698–1707.

NOBLE, A. J., and STAGG, S. M. 2015. Automated batch fiducial-less tilt-series alignment in Appion using Protomo. *J Struct Biol*, **192**, 270–278.

Noda, T., Ebihara, H., Muramoto, Y., Fujii, K., Takada, A., Sagara, H., Kim, J. H., Kida, H., Feldmann, H., and Kawaoka, Y. 2006. Assembly and budding of Ebolavirus. *PLoS Pathog*, **2**, e99.

Paul-Gilloteaux, P., Heiligenstein, X., Belle, M., Domart, M. C., Larijani, B., Collinson, L., Raposo, G., and Salamero, J. 2017. eC-CLEM: flexible multidimensional registration software for correlative microscopies. *Nat Methods*, **14**, 102–103.

Pfeffer, S., Woellhaf, M. W., Herrmann, J. M., and Forster, F. 2015. Organization of the mitochondrial translation machinery studied in situ by cryoelectron tomography. *Nat Commun*, **6**, 6019.

Picco, A., Irastorza-Azcarate, I., Specht, T., Boke, D., Pazos, I., Rivier-Cordey, A. S., Devos, D. P., Kaksonen, M., and Gallego, O. 2017. The In Vivo Architecture of the Exocyst Provides Structural Basis for Exocytosis. *Cell*, **168**, 400–412 e18.

Pierson, J., Fernandez, J. J., Bos, E., Amini, S., Gnaegi, H., Vos, M., Bel, B., Adolfsen, F., Carrascosa, J. L., and Peters, P. J. 2010. Improving the technique of vitreous cryo-sectioning for cryo-electron tomography: electrostatic charging for section attachment and implementation of an anti-contamination glove box. *J Struct Biol*, **169**, 219–225.

Pierson, J., Ziese, U., Sani, M., and Peters, P. J. 2011. Exploring vitreous cryo-section-induced compression at the macromolecular level using electron cryo-tomography; 80S yeast ribosomes appear unaffected. *J Struct Biol*, **173**, 345–349.

Rickgauer, J. P., Grigorieff, N., and Denk, W. 2017. Single-protein detection in crowded molecular environments in cryo-EM images. *Elife*, 6.

Rigort, A., Bauerlein, F. J., Leis, A., Gruska, M., Hoffmann, C., Laugks, T., Bohm, U., Eibauer, M., Gnaegi, H., Baumeister, W., and Plitzko, J. M. 2010. Micromachining tools and correlative approaches for cellular cryo-electron tomography. *J Struct Biol*, **172**, 169–279.

Salje, J., Zuber, B., and Löwe, J. 2009. Electron cryomicroscopy of E. coli reveals filament bundles involved in plasmid DNA segregation. *Science*, **323**, 509–512.

Sandblad, L., Busch, K. E., Tittmann, P., Gross, H., Brunner, D., and Hoenger, A. 2006. The Schizosaccharomyces pombe EB1 homolog Mal3p binds and stabilizes the microtubule lattice seam. *Cell*, **127**, 1415–1424.

Sartori, A., Gatz, R., Beck, F., Rigort, A., Baumeister, W., and Plitzko, J. M. 2007. Correlative microscopy: bridging the gap between fluorescence light microscopy and cryo-electron tomography. *J Struct Biol*, **160**, 135–145.

Sartori Blanc, N., Studer, D., Ruhl, K., and Dubochet, J. 1998. Electron beam-induced changes in vitreous sections of biological samples. *J Microsc*, **192**, 194–201.

Scheffer, M. P., Eltsov, M., and Frangakis, A. S. 2011. Evidence for short-range helical order in the 30-nm chromatin fibers of erythrocyte nuclei. *Proc Natl Acad Sci U S A*, **108**, 16992–16997.

Schellenberger, P., Kaufmann, R., Siebert, C. A., Hagen, C., Wodrich, H., and Grunewald, K. 2014. High-precision correlative fluorescence and electron cryo microscopy using two independent alignment markers. *Ultramicroscopy*, **143**, 41–51.

Schorb, M., and Briggs, J. A. 2014. Correlated cryo-fluorescence and cryo-electron microscopy with high spatial precision and improved sensitivity. *Ultramicroscopy*, **143**, 24–32.

Schorb, M., Gaechter, L., Avinoam, O., Sieckmann, F., Clarke, M., Bebeacua, C., Bykov, Y. S., Sonnen, A. F., Lihl, R., and Briggs, J. A. 2017. New hardware and workflows for semi-automated correlative cryo-fluorescence and cryo-electron microscopy/tomography. *J Struct Biol*, **197**, 83–93.

Schwartz, C. L., Sarbash, V. I., Ataullakhanov, F. I., Mcintosh, J. R., and Nicastro, D. 2007. Cryo-fluorescence microscopy facilitates correlations between light and cryo-electron microscopy and reduces the rate of photobleaching. *J Microsc*, **227**, 98–109.

Sorzano, C. O., Messaoudi, C., Eibauer, M., Bilbao-Castro, J. R., Hegerl, R., Nickell, S., Marco, S., and Carazo, J. M. 2009. Marker-free image registration of electron tomography tilt-series. *BMC Bioinformatics*, **10**, 124.

Szwedziak, P., Wang, Q., Bharat, T. A., Tsim, M., and Lowe, J. 2014. Architecture of the ring formed by the tubulin homologue FtsZ in bacterial cell division. *Elife*, **4**, e04601.

Van Driel, L. F., Valentijn, J. A., Valentijn, K. M., Koning, R. I., and Koster, A. J. 2009. Tools for correlative cryo-fluorescence microscopy and cryo-electron tomography applied to whole mitochondria in human endothelial cells. *Eur J Cell Biol*, **88**, 669–684.

Villa, E., Schaffer, M., Plitzko, J. M., and Baumeister, W. 2013. Opening windows into the cell: focused-ion-beam milling for cryo-electron tomography. *Curr Opin Struct Biol*, **23**, 771–777.

Wolff, G., Hagen, C., Grunewald, K., and Kaufmann, R. 2016. Towards correlative super-resolution fluorescence and electron cryo-microscopy. *Biol Cell*, **108**, 245–258.

Zuber, B., Nikonenko, I., Klauser, P., Muller, D., and Dubochet, J. 2005. The mammalian central nervous synaptic cleft contains a high density of periodically organized complexes. *Proc Natl Acad Sci U S A*, **102**, 19192–19197.

9

Correlative Cryo Soft X-ray Imaging

Eva Pereiro[1], Francisco Javier Chichón[2], and Jose L. Carrascosa[2]

[1] *Mistral beamline, ALBA Light Source, Cerdanyola del Vallès, Barcelona, Spain*
[2] *Department of Macromolecular Structures, Centro Nacional de Biotecnologia (CNB-CSIC), Madrid, Spain*

9.1 Introduction to Cryo Soft X-ray Microscopy

Structural cell biology demands detailed structural and functional descriptions of the different cellular components which must be correlated with a topological 3D map of these components at the whole cellular level. Optical microscopy has a privileged position in this field due to its availability at laboratories worldwide and the possibility to image live. The use of fluorescent dyes to label specific proteins or organelles, together with the implementation of high-resolution approaches, has revolutionized the technique, and nowadays distances between labeled features can be determined well beyond the diffraction limit [1]. However, only the tagged features are visualized, therefore producing a limited picture of the system. On the other hand, transmission electron microscopy (TEM) allows cell visualization at a higher spatial resolution, thus revealing ultrastructural cellular details. But this technique requires vacuum and cellular sectioning as the multiple scattering of electrons imposes a restrictive sample thickness limitation. Thus, sample preparation for TEM requires complex protocols, mainly based on fixation, staining, dehydration, resin embedding and sectioning. The use of mechanical means for producing sections of thick samples results in many important drawbacks and artifacts, such as marks, compression, deformation etc. More recently, cryo-fixation [2-4] has been used to avoid dehydration and staining allowing visualizing the sample in its near-native state. Overall, whole cell visualization by TEM is complicated as stacked 2D sections are required, resulting in difficult and time-consuming experiments.

The possibility to use scanning electron microscopy (SEM) to study resin-embedded samples has led to the implementation of a number of approaches, such as serial block face imaging techniques that use either ion beam milling, or an in situ ultramicrotome with a diamond knife, to progressively expose the sample volume to SEM visualization. In these cases, the information is mainly limited by the use of chemical fixation and multiple layers of staining with heavy metals, used to make the sample conductive. Although it is a destructive approach, it has been successfully applied for studying larger volumes, for example in the study of neural tissue nanostructure [5]. In the case of vitrified samples, a focused ion beam can be used for 3D imaging. This procedure has a long tradition in materials science, but it has only recently been incorporated for biological samples [6]. There are two main applications of ion milling of a thick biological sample: integration with a scanning microscope to sequentially image through the sample volume [7], and the use of the milling capabilities to produce accurate sections of the sample with thicknesses appropriate for transmission electron microscopy [8]. Although the technical demands for this type of sample production preclude its use as a standard, mid-throughput technique, it has shown an impressive potential in sample preparation for cryo-electron tomography [9]. Nevertheless, all of these ex situ and integrated sectioning techniques are prone to artifacts.

Even though the above-mentioned visible light and electron microscopy techniques are broadly used, the correlation of information they provide is hindered firstly by the existing resolution gap between them, and secondly by the different sample thickness range each of them can handle. In this frame, an emerging technique such as cryo soft X-ray tomography (cryo-SXT) [10] can provide structural information at the level of a whole cell without further sample preparation except for the cryo-fixation required to prevent radiation damage while collecting the data. The penetration power of soft X-rays in the so-called *water window* spectral range, between the inner-shell absorption edges of carbon and oxygen (from 284 eV to 543 eV), allows penetrating water layers of up to 10 μm thickness (Figure 9.1a) while carbon-rich structures are visualized with good absorption contrast [11] (Figure 9.1b). Thus, frozen-hydrated specimens can be imaged close to their native state providing significant complementary information to existing biological imaging techniques at a spatial resolution of 30 nm.

Efforts in the field of X-ray microscopy started slowly around 1975. Two major steps have made cryo-SXT a useful imaging technique for cell biology: the development of bright, high-intensity, energy tunable synchrotron sources and the nanofabrication techniques necessary for specific optical lens development [12, 13]. Indeed, the manufacturing progress in diffractive Fresnel Zone Plate (FZP) lenses has been a key issue as

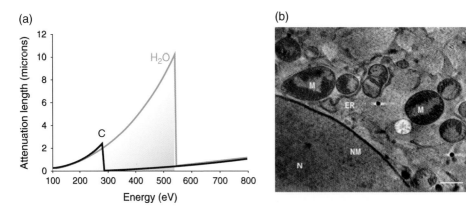

Figure 9.1 (a) attenuation length showing the water window energy range (284 eV–543 eV). Several microns of water can be penetrated just below the oxygen K edge; (b) reconstructed slice of a vitrified hepatocyte (courtesy of AJ Pérez-Berná) at 520 eV showing the natural contrast of the carbon-based structures of the cell (N nucleus, NM nuclear membrane, ER endoplasmic reticulum, M mitochondria). Scale bar: 1 μm.

the overall achievable resolution is governed by the lens (assuming minimal aberrations, proper illumination, stable operation, ideal sample and CCD detector). Finally, the development of cryogenic systems to handle vitrified samples in X-ray microscopes has made cellular X-ray imaging possible. In what follows, we will focus our attention on the tomographic capabilities of transmission soft X-ray microscopes (TXM) in the field of three-dimensional (3D) biological imaging. Other microscope types are also available at synchrotron sources, such as for instance scanning transmission X-ray microscopes that can provide useful information such as 3D cryo X-ray fluorescence maps [14]. For a full review of direct-image-forming X-ray microscopes, please refer to Sakdinawat & Attwood [15].

As up today, cryo SXT has been used for imaging a variety of biological samples ranging from isolated virus [16], to yeasts and protozoa [17-19], and mammalian cells [20-32]. It has proven to be a useful imaging tool for investigating pathogen-host interaction [21, 31] (Figure 9.2), cellular contacts [26], cellular organelles [27], intracellular nanoparticle internalization [29] and the nucleus [33, 34]. As an example, Figure 9.2 illustrates the capabilities of cryo-SXT to investigate the cytoplasm modifications of Hepatitis C virus (HCV) replicon-bearing cells [31].

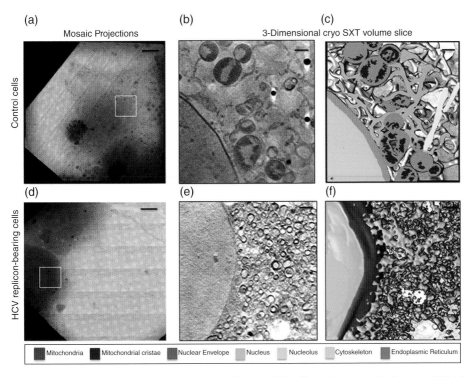

Figure 9.2 **3D reconstructions of whole control and HCV replicon-bearing cells by cryo SXT.** (a) Mosaic projections of the control cells grid mesh. White square indicates the cryo SXT field of view of the tomography reconstruction shown in (b); (b) volume slice of control cell; (c) segmentation of (b). (d) Mosaic projections of HCV replicon-bearing cell. White square indicates the cryo SXT field of view of the tomography reconstruction shown in (e); (e) volume slice of HCV replicon-bearing cell; (f) segmentation of (e). Scale bars: (a) and (d) 10 μm, (b, c) and (e, f) 1 μm. Courtesy of AJ Pérez-Berná.

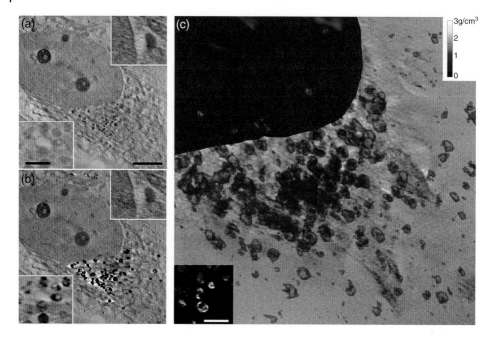

Figure 9.3 **Reconstructed slices of MCF-7 cell incubated with superparamagnetic nanoparticles (DMSA-Fe₂O₃).** (a) Cellular volume reconstructed at 700 eV below the Fe L_3 edge; (b) Cellular volume reconstructed at the Fe L_3 edge 709 eV. As seen in the figure, the carbon cellular structures have the same absorption (upper-right insets), while the iron oxide nanoparticles show higher absorption at the L_3 absorption edge (bottom-left insets); (c) segmentation of the iron oxide densities and the cell nucleus in blue (Conesa *et al.* 2016). Scale bars: (a, b) 3 μm, insets in (a, b, c) 500 nm. Courtesy of J.J. Conesa.

Furthermore, one of the distinctive features of X-rays is their ability to be absorbed differentially by the elements constituting the sample. Thus, in addition to the capabilities of SXT to reveal structural information, chemical specificity can also be exploited to create 2D or 3D differential contrast maps [35–37], below and above an interesting absorption edge of a particular element, for which the attenuation of the photons varies rapidly due to the photoelectric absorption of each element (set by the electron binding energies). In this case, the use of cryo-fixed thick samples for soft X-ray spectro-microscopy would be restricted to the *water window* energy range, for instance to study the biomineralization process [37] at the Ca L edges (around 350 eV). However, other sample fixation methods for which water is eliminated from the cell (such as critical point sublimation or freeze drying) allows using a wider energy range for reaching interesting absorption edges. For instance, Conesa and collaborators were interested in the mass quantification of superparamagnetic nanoparticles in cancer cells related to their use in nano-biomedicine for diagnosis, drug-delivery and hyperthermia treatment [36]. Figure 9.3 shows how using the iron L_3 edge (709 eV), the accumulation of these nanoparticles along the endocytic pathway of the MCF-7 cell is highlighted and the 3D mass quantification per cell related to its uptake time can be determined with a sensitivity of one nanoparticle within the cellular volume.

Finally, the spatial resolution achievable with a TXM is set by the FZP objective lens. This lens is a circular diffraction grating whose concentric zones, with radially decreasing

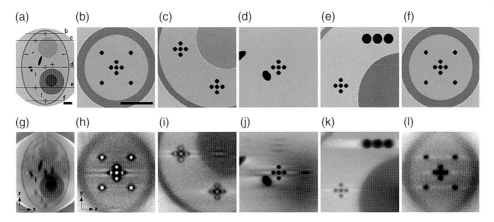

Figure 9.4 **Comparison of reconstructed slices of a simulated 9 µm thick pseudo *candida albicans* cell imaged with a 40 nm FZP lens in a totally incoherent situation** (numerical apertures of condenser and objective lenses match: 48.8 nm resolution, 2.63 µm DOF): phantom (first row) and standard reconstruction (second row); (a, g) *x-z* planes where *x-y* slices (b, f and h, l) at different *z*-positions in (a) and their corresponding reconstruction are shown: −3.75 µm (b, h), −3 µm (c, i), 0 µm (d, j), 2.25 µm (e, k) and 3.75 µm (f, l). Scale bars = 1 µm. Courtesy of J. Otón.

zone widths, are designed to focus light and form a real image [38]. The spatial resolution of the lens is given by the width of the last opaque zone, the so-called outermost zone width Δr_N, therefore the smaller the width the higher the spatial resolution achievable. FZP lenses providing up to 15 nm resolution in 2D, have been successfully manufactured [39]. However, this high resolution comes at the price of a smaller depth of field (DoF= $\pm\ 2\Delta r_N^2/\lambda$, where λ is the wavelength), described as the distance for which 20% of the maximum axial intensity is lost [40], which in practice limits the thickness of the sample in focus. This is a limitation factor when imaging samples thicker than the DoF, which in practice is often the case. Figure 9.4 shows the effect of the DoF on a 9 µm thick phantom cell imaged with a 40 nm FZP lens (courtesy of J. Otón). A discussion on the depth of field limitation is outside the scope of this chapter, but details on this subject can be found in [38, 41–44]. We would like to highlight that new data collection methods and algorithms are being developed to extend the DoF limitation of the lens in cryo-SXT [45]. Nowadays, the most routinely used FZP lenses for imaging whole vitrified cells have outermost zone widths (Δr_N) of 40 nm and 25 nm, corresponding to theoretical DoF of approximately 3 µm and 1 µm, respectively. For a detailed explanation of the principles of FZP X-ray microscopes please refer to [38, 42].

9.2 Cryo-SXT Correlation with Visible Light Microscopy

Functional studies are achievable now by correlating visible light fluorescence information prior and after vitrification with the X-ray tomogram ultrastructures. Finder grids are normally very useful, allowing the easy location of the same region in images showing very different signals. A usual way to proceed is the following. Fluorescence optical microscopy is used before vitrification to enable tracking relevant components or events within live cells. Once the samples are vitrified and loaded into the TXM chamber, the relevant biological features are localized again with an on-line fluorescence microscope allowing tracking the interesting target that should be imaged with X-rays (Figure 9.5). The capability to generate cryo-fluorescence images inside the soft X-ray microscope has represented a great advantage for correlative purposes, but checking the quality of the samples before introducing them into the microscope vacuum chamber is mandatory for successful SXT experiments as grids can get damaged and/or cells can move during the process [21, 23, 27, 31, 46–48]. Further development of compatible sample holders for all instruments is necessary to facilitate sample transfer between instruments while minimizing risk and possible contamination. Autogrids such as those used in electron microscopy, for instance, can be a useful tool toward this objective.

Recently, a correlative study of STORM and cryo-SXT has been published highlighting the possibility of specific molecular identification in combination with high-resolution 3D cellular information [32]. Varsano and collaborators were interested in localizing the early formation of cholesterol crystals in relation with atherosclerosis using macrophage cell cultures enriched with low density lipoproteins. By combining specific markers on the desired epitopes revealed by STORM and cryo-SXT cellular volumes, early crystals and 2D cholesterol domains were identified on the plasma membrane and in intracellular locations (see Figure 9.6). We believe the combination of super-resolution fluorescence techniques and cryo-SXT is an extremely powerful approach and more examples of such correlative studies are expected in the near future.

9.3 Cryo-SXT Correlation with Cryo X-ray Fluorescence

Another combination of information that should be highlighted here is cryo-SXT and cryo X-ray fluorescence (cryo-XRF). Actually, nano beams have started to be available at several synchrotron facilities which, in addition, have also developed cryogenic

Figure 9.5 Correlative workflow for visible light fluorescence and cryo SXT. Upper panel shows the schematic workflow. (a) Visible light image (MitoTracker red) in the ocular of the cryo-fluorescence microscope for a first inspection of the grid quality. Yellow square encloses the region of the grid imaged by cryo-SXT. (b) Red channel mosaic of the grid imaged by cryo-fluorescence off-line. (c) Wide field mosaic of the grid imaged in cryo-fluorescence microscope "on-line" installed in the soft X-ray transmission microscope. This image is used for coordinate registration and as a second visual inspection of the grid quality. (d) Soft X-ray mosaic of the grid square labeled with a yellow square in *a, b* and *c*. This mosaic is used for a fine coordinate registration of the "region of interest" (ROI, yellow square) and for correlation with fluorescence images (*in-vivo, off-line, on-line*). It allows also vitrification assessment (ice quality and thickness). (e) Cryo-fluorescence image (red and green channels, mitochondria, and GFP-labeled protein respectively) of the same square in d. (f) Fluorescence and X-ray mosaic overlay of the ROI of the cell: This correlation allows the user to locate the biologically relevant features as the cell is generally bigger than the X-ray field of view (yellow square). (g) Cryo soft X-ray tilt series acquisition scheme. Before the automatic acquisition, all the parameters must be settled (maximum tilt range, tilt step, exposure time). (h) Cryo soft X-ray tomographic reconstruction (preprocessing, tilt-series alignment, and 3D reconstruction) require a computer and dedicated software. Then volume inspection, segmentation, and analysis (qualitative or quantitative), as well as the correlation again with the fluorescent signal (2D) will also require dedicated software.

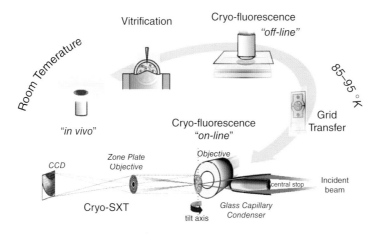

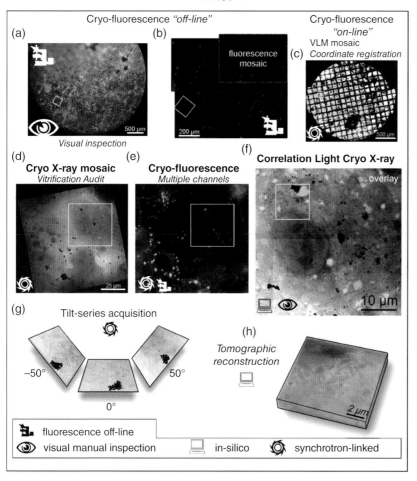

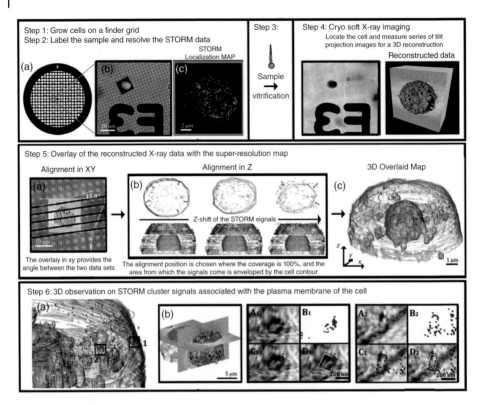

Figure 9.6 **Correlative workflow of STORM and cryo SXT**. Step 1: Cells were grown on gold finder grids with fiducial markers that allow navigation to desired locations on the grid. The cells were incubated for 48 h with acetylated low density lipoproteins. Step 2: Cells were fixed and incubated with primary and secondary antibodies. The super-resolution fluorescence signal was resolved by STORM. Step 3: Grids were vitrified. Step 4: X-ray tomograms were taken of the same cells that were analyzed by STORM. Data were reconstructed (field of view 18 μm^3). Step 5: Overlay of the data in x, y and z. Step 6: (a) 3D reconstruction of the same cell from a side view. The localization map of the corresponding resolved super-resolution image (red spots) is 1 μm in thickness and is located on the upper part of the cell, above the nucleus. Black rectangles indicate clusters of STORM signals. (b) Side orientation of the superimposed data combined with a perpendicular slice. (C1-F1) and (C2-F2) show magnifications of the crystals indicated by 1 and 2 in (A) viewed in SXT, in STORM and superimposed, respectively. Courtesy of N. Varsano and Prof. L. Addadi.

capabilities therefore allowing collecting cryo-XRF data at spatial resolutions from 100 nm down to 30 nm matching quite well the resolution of cryo-SXT. A latest example of this by Kapishnikov et al. is the quantification of Fe bound either to hemoglobin or to hemozoin crystals in red blood cells infected by *Plasmodium falciparum* [49] (Figure 9.7). A key drug target for malaria has been the crystallization pathway of the iron-containing molecule heme, which is the toxic byproduct of hemoglobin digestion. How heme is crystallized remains unclear, but current models predict very different crystallization rates. To shed some light on the different possible models, Kapishnikov and collaborators estimated the *in-vivo* rate of heme crystallization in the malaria parasite, by measuring element-specific concentrations at defined locations in infected red blood cells combining cryo-SXT to obtain the 3D parasite ultrastructure with X-ray fluorescence microscopy to

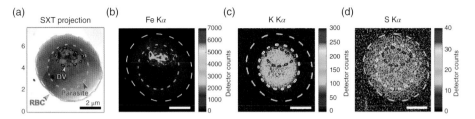

Figure 9.7 **Maps of the trophozoite stage of the malaria parasite**. (a) cryo-SXT projection at 520 eV of a red blood cell (RBC) infected with *Plasmodium falsciparum* in which the parasite and the digestive vacuole (DV) containing the hemozoine crystals have been indicated. (b, c, and d) iron, potassium, and sulfur X-ray fluorescence maps, respectively, of the same cell. Scale bars: 2μm. Courtesy of S. Kapishnikov.

measure specific iron concentrations. Their results are in agreement with a crystallization occurring via the heme detoxification protein [50].

The combination of 3D structural data from cryo-SXT and 2D or 3D X-ray fluorescence data of specific elements is certainly an interesting path for tackling many complex biochemical problems.

9.4   Cryo-SXT Correlation with TEM

A very interesting level of correlation would be to locate specific structures in the cell globally with cryo-SXT and then to visualize them at higher magnification and resolution with TEM. In contrast to visible light fluorescence microscopy, where the very different image signal types make the image alignment with TEM or SXT difficult, the correlation between the projections or volumes coming from soft X-rays and electrons is facilitated by the fact that both techniques are absorption-based and thus present a similar view [21]. Nevertheless, there are several features which complicate this combination. The first one is radiation damage as the signal-to-noise required for a well resolved tomogram using both soft X-rays and electrons is significant. If cryo-SXT should be combined with TEM, the dose would have to be spread between the two techniques in order to stay within the dose limit [51]. Therefore, a lower signal-to-noise ratio per projection as well as, for instance, a larger r angular step in the tomography acquisition schemes would be required for this approach. The second problem is related to the sample thickness. While a 5 microns thick-cell is suitable for visible light microscopy and cryo-SXT, TEM requirements reduce the visualization to a much thinner part of the cell (as, for instance, the cytoplasm periphery), making correlative attempts quite limited. This problem can be overcome by sectioning or trimming the sample after visualization by cryo-SXT. Appropriate workflows would then need to be developed to make correlative cryo-SXT and TEM useful. One possibility, already pointed out in [21], is based in the freeze-substitution of the grid after visualization by cryo-SXT. This technique allows embedding cryo-preserved samples in an acrylic resin at low temperature [52]. Once the grid is embedded as a block, it would need to be trimmed to locate the specific areas imaged in cryo-SXT (Figure 9.8). However, this method implies a manual, risky and time consuming manipulation with low success

VLM-SXM-mosaic overlay   TEM section

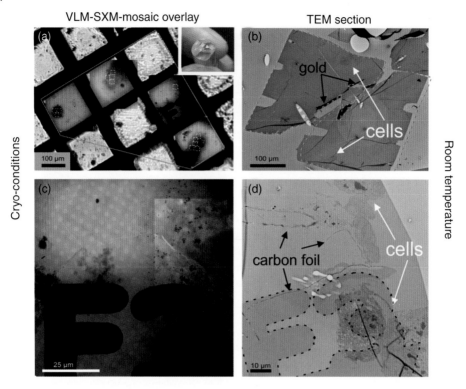

Figure 9.8 **Cryo-SXM-TEM correlation.** (a) Overlay between visible light microscopy (VLM) and four soft X-ray mosaics (SXM) with interesting cell areas for which cryo-SXT was acquired (white squares). This overlay was used to locate the region of the grid to be trimmed in the ultramicrotome (orange area). Inset shows the embedded sample. (b) Grid square SXM with visible light fluorescence image overlay of a specific cell. (c) and (d) Embedded TEM-section images corresponding to regions in (a) and (b) respectively.

rate, mainly due to the non-planar mechanical sectioning and the deformations that this process produces. Also, the sections are normally fragmented and require very careful handling.

One of the problems found along this correlative workflow was that the grid foil from irradiated areas after cryo-SXT broke more easily than those from non-irradiated areas after the dehydration required in the freeze substitution process. This fragility of the carbon foil, on which the cells are grown or deposited, makes the correlation process extremely complicated. This problem could be partially overcome using non-mechanical trimming and a "slice-and-view" approach such as in FIB-SEM. SEM would in addition allow for an easier correlation before starting the FIB milling. The generation of thick lamellas (200–300 nm) for electron tomography would then be more accurate as precise three-dimensional information given by cryo-SXT would be available. Furthermore, this equipment could be used in cryo-conditions to avoid sample damage due to dehydration as in [53], achieving similar resolution as in SXT or as in cryo-CLEM [54] by removing non desired areas to create an oblique lamella for cryo-electron tomography. In that sense, the data from cryo-SXT could be used mainly to find the particular area of interest within the cellular volume and, therefore, the dose could be reduced. Then cryo-CLEM workflows involving cryo FIB milling could be followed in a much more accurate and reliable way accounting for the depth information given by cryo-SXT.

9.5   Multiple Correlation and Integration of Methods

The above-mentioned examples highlight the central role of cryo-SXT as an ideal scaffold for building together pieces of information obtained from methods with large differences in resolution and sample thickness capabilities. Morphological data from light microscopy, super-resolution fluorescence information, and high-resolution structural features derived from electron microscopy, together with chemical information from different spectroscopic sources, can be integrated into the structural framework provided by cryo-SXT. It is clear that the successful development of this approach to fully understand the cellular complexity at nanometric resolution would depend very much on the implementation of methods to integrate the data from these different sources. Working at the interface of these approaches demands addressing different aspects: from the data format and annotated description to facilitate integration at correlative resolutions, up to the development of flexible sample holders allowing the analysis of the very same samples in different experimental setups. Several ongoing efforts on the development of multi-cryotransfer systems are quite promising at this level, which are aligned within the current development of different integrated microscopy equipment (CLEM, ILEM, SECOM, ASEM, etc. [55–57]).

On the other hand, one of the more attractive aspects of cryo-SXT as an integrative way to build complex cellular models at correlative resolution levels, is the possibility to use dedicated probes which could be tracked using different methods. The existence of these internal markers will then be instrumental for building the integration of data into the SXT volume. There are many efforts in this area, as for example to incorporate quantum dot based probes to correlate TEM and SXT, as well as the use of different metal-labelled proteins which could be detected by electron density at TEM and also provide enhanced detection by near-edge X-ray spectral imaging [36]. Recombinant proteins tagged with fluorescent peptides [58], together with the incorporation of self-labeling enzymes using DAB oxidation [59] are examples to illustrate the correlation of fluorescence, high-resolution light microscopy and TEM detection. Extension of the use of this type of tags towards cryo-SXT is a possible way to provide truly correlative information across a wide range of resolution levels.

Acknowledgments

We thank Dr. A.J. Pérez-Berná, Dr. J.J. Conesa, Dr. J. Otón, Dr. S. Kapishnikov, N. Varsano, Prof. L. Addadi for their essential figure contributions, which are the backbone of this chapter. The cryo-SXT works shown have all been supported by ALBA Light Source standard proposals and have been collected at the Mistral beamline. This work was partially supported by a grant from the Spanish Ministerio de Economia y Competitividad (BFU2014-54181) to JLC.

References

1 Hell SW. Far-field optical nanoscopy. *Science* 2007 **316**:1153–58.

2 Dubochet J, Adrian M, Chang JJ, Homo JC, Lepault J, McDowall AW et al. Cryo-electron microscopy of vitrified specimens. *Quarterly Reviews of Biophysics* 1988 **21**(2):129–228.

3 Dubochet J. Vitreous Water. *Handbook of Cryo-Preparation Methods for Electron Microscopy*, Boca Raton, FL: CRC Press; 2008: 3–14.

4 Rigort A, Bräuerlein FJB, Leis A, Gruska M, Hoffmann C, Laugks T, et al. Micromachining tools and correlative approaches for cellular cryo-electron tomography. *Journal of Structural Biology* 2010 **172**:169–79.

5 Denk W, and Horstmann H. Serial blockface scanning electron microscopy to reconstruct three-dimensional tissue nanostructure. *PLoS Biol* 2004 **2**(11):e329.

6 Narayan K, and Subramaniam S. Focused ion beams in biology. *Nat Meth* 2015 **12**(11):1021–31.

7 Heymann JA, Shi D, Kim S, Bliss D, Milne JLS, Subramaniam S. 3D imaging of mammalian cells with ion-abrasion scanning electron microscopy. *J Struct Biol* 2009 **166**(1): 1–7.

8 Marko M, Hsieh C, Moberlychan W, Mannella CA, Frank J. Focused ion beam milling of vitreous water: prospects for an alternative to cryo-ultramicrotomy of frozen-hydrated biological samples. *J Microsc* 2006 **222**(Pt 1):42–47.

9 Rigort A, and Plitzko JM. Cryo-focused-ion-beam applications in structural biology. *Arch Biochem Biophys* 2015 **581**:122–30. Sartori, A., R. Gatz, et al. Correlative microscopy: bridging the gap between fluorescence light microscopy and cryo-electron tomography. *J Struct Biol* 2007 160(2):135–45.

10 Schneider G, Guttmann P, Heim S, Rehbein S, Mueller F, Nagashima K, et al. Three-dimensional cellular ultrastructure resolved by X-ray microscopy. *Nature Methods* 2010 **10**:1–3.

11 Wolter H. Spiegelsysteme streifenden Einfalls als abbildende Optiken für Röntgenstrahlen. *Annalen der Physik* 1952 **10**:94–114.

12 Schmahl G, Rudolph D, Guttmann P, Christ O. Zone plates for X-ray microscopy pp:63–74. In Schmahl G and Rudolph D (eds.) X-ray microscopy, vol. **43**. Springer Series in Optical Sciences; 1984.

13 Michette AG, Morrison GR and Buckley CJ. (eds.) X-ray microscopy III. Springer Series in Optical Sciences. Berlin: Springer-Verlag; 1992.

14 Yuan Y, Chen S, Paunesku T, Gleber SC, Liu WC, Doty CB, et al. Epidermal growth factor receptor targeted nuclear delivery and high-resolution whole cell X-ray imaging of $Fe_3O_4@TiO_2$ nanoparticles in cancer cells. *ACS Nano* 2013 **7**(12):10502–517.

15 Sakdinawat A, and Atwood D Nanoscale X-ray imaging. *Nature Photonics* 2010 **4**:840–848.

16 Carrascosa JL, Chichon FJ, Pereiro E, Rodriguez MJ, Fernandez JJ, Esteban M, et al. Cryo X-ray tomography of vaccinia virus membranes and inner compartments. *Journal of Structural Biology* 2009 **168**:234–39.

17 Parkinson DY, McDermott G, Etkin LD, Le Gros MA, Larabell CA. Quantitative 3D imaging of eukaryotic cells using soft X-ray tomography. *Journal of Structural Biology* 2008 **162**: 380–86.

18 Uchida M, McDermott D, Wetzler M, Le Gros M, Myllys M, Knoechel C Soft X-ray tomography of phenotypic switching and the cellular response to antifungal peptoidsin Candida albicans. *PNAS* 2009 **106**:19375–80.

19 Hummel EP, Guttmann S, Werner B, Tarek G, Schneider M, Kunz AS, et al. 3D Ultrastructural organization of whole Chlamydomonas reinhardtii cells studied by nanoscale soft x-ray tomography. *PLOS ONE* 2012 **7**(12):e53293.

20 Hanssen E, Knoeckel C, Klonis N, Abur-Bakar N, Deed S, Le Gros M, et al. Cryo transmission X-ray imaging of the malaria parasite P. falciparum. *Journal Structural Biology* 2011 **173**:161–68.

21 Chichon FJ, Rodriguez MJ, Pereiro E, Chiappi M, Perdiguero B, Guttmann P, et al. Cryo nano-tomography of vaccinia virus infected cells. *Journal of Structural Biology* 2012 **177**:202–11.

22 Clowncy EJ, Lc Gros MA, Mosley CP, Clowney FG, Markenskoff-Papadimitriou EC, Myllys M, et al. Nuclear aggregation of olfactory receptor genes governs their monogenic expression. *Cell* 2012 **151**:724–37.

23 Hagen C, Guttmann P, Klupp B, Werner S, Rehbein S, Mettenleiter T, et al. Correlative VIS-fluorescence and soft X-ray cryo-microscopy/tomography of adherent cells. *Journal of Structural Biology* 2012 **177**:193–201.

24 Hanssen E, Knoechel C, Dearnley M, Dixon MWA, Le Gros M, Larabell C, et al. Soft X-ray microscopy analysis of cell volumen and hemoglobina content in erythrocytes infected with asexual and sexual stages of *Plasmodium falciparum. Journal Structural Biology* 2012 **177**:224–32.

25 Kapishnikov S, Weiner A, Shimoni E, Guttmann P, Schneider G, Dahan-Pasternak N, et al. Oriented nucleation of hemozoin at the digestive vacuole membrane in *Plasmodium falsiparum. PNAS* 2012 **109**:11188–93.

26 Cruz-Adalia A, Calabia-Linares C, Ramirez-Santiago G, Torres-Torresano M, Feo L, Galán-Díez M, et al. T cells capture and destroy bacteria by trans-infection from dendritic cells. *Cell Host & Microbe* 2014 **15**:611–22.

27 Duke EMH, Razi M, Weston A, Guttmann P, Werner S, Henzler K, et al. Imaging endosomes and autophagosomes in whole mammalian cells using correlative cryo-fluorescence and cryo-soft X-ray microscopy (cryo-CLXM). *Ultramicroscopy* 2014 **143**:77–87.

28 Chen HY, Chiang DML, Lin ZJ, Hsieh CC, Yin GC, Weng IC, et al. Nanoimaging granule dynamics and subcellular structures in activated mast cells using soft X-ray tomography. *Scientific Report* 2016 **6**:34879.

29 Chiappi M, Conesa JJ, Pereiro E, Sánchez Sorzano CO, Rodríguez MJ, Henzler K, et al. Cryo-soft X-ray tomography as a quantitative three-dimensional tool to model nanoparticle:cell interaction. *J. of Nanobiotechnology* 2016 **14**:15.

30 LeGros MA, Clowney EJ, Magklara A, Yen A, Markenscoff-Papadimitrou E, Colquitt B, et al. Soft X-ray tomography reveals gradual chromatin compaction and reorganization during neurogenesis in vivo. *Cell Reports* 2016 **17**:2125–36.

31 Pérez-Berná AJ, Rodríguez MJ, Chichón FJ, Friesland MF, Sorrentino A, Carrascosa JL, et al. Structural changes in cells images by soft X-ray cryo-tomography during Hepatitis C virus infection. *ACS Nano* 2016 **10**:6597–611.

32 Varsano N, Dadosh T, Kapishnikov S, Pereiro E, Shimoni E, Jin X, et al. Development of correlative cryo-soft X-ray tomography and stochastic reconstruction microscopy. A study of cholesterol crystal early formation in cells. *JACS* 2016 **138**:14931–40.

33 Hagen C, Werner S, Carregal-Romero S, Malhas AN, Klupp BG, Guttmann P, et al. Multimodal nanoparticles as alignment and correlation markers in fluorescence/soft X-ray cryo-microscopy/tomography of nucleoplasmic reticulum and apoptosis in mammalian cells. *Ultramicroscopy* 2014 **146**:46–54.

34 Hagen C, Dent KC, Zeev-Ben-Mordehai T, Grange M, Bosse JB, Whittle C, et al. Structural basis of vesicle formation at the inner nuclear membrane. *Cell* 2015 **163**:1692–701.

35 Zhang X, Balhorn R, Mazrimas J, Kirtz J. Mapping and measuring DNA to protein ratios in mammalian sperm head by XANES imaging. *Journal of Structural Biology* 1996 **116**:335–44.

36 Conesa JJ, Otón J, Chiappi M, Carazo JM, Pereiro E, Chichón FJ, et al. Intracellular nanoparticles mass quantification by near-edge absorption soft X-ray nanotomography. *Scientific Reports* 2016 **6**:22354.

37 Sviben S, Gal A, Hood MA, Bertinetti L, Politi Y, Bennet M, et al. A vacuole-like compartment concentrates a disordered calcium phase in a key coccolithophorid alga. *Nature Com.* 2016 7:11228.

38 Attwood D. *Soft X-rays and Extreme Ultraviolet Radiation.* Principles and Applications. Cambridge: Cambridge University Press; 2000.

39 Chao W, Harteneck BD, Liddle JA, Anderson EH, Attwood DT. Soft X-ray Microscopy at a spatial resolution better than 15 nm. *Nature* 2005 **435**:1210–13.

40 Born M, Wolf E. *Principles of optics: electromagnetic theory of propagation, interference and diffraction of light.* Cambridge: Cambridge University Press; 1999.

41 Weiss D, Schneider G, Niemann B, Guttmann P, Rudolph D, Schmahl G Computed tomography of cryogenic biological specimens based on X-ray microscopic images. *Ultramicroscopy* 2000 **84**:185–97.

42 Howells M, Jacobsen C, Warwick T. *Principles and Applications of Zone Plate X-ray Microscopes, chapter 13, Science of Microscopy.* New York: Springer; 2007.

43 Bertilson M, von Hofsten O, Hertz HM, Vogt U Numerical model for tomographic image formation in transmission X-ray microscopy. *Optics Express* 2011 **19**: 11578–83.

44 Oton J, Sorzano COS, Pereiro E, Cuenca-Alba J, Navarro R, Carazo JM, et al. Image formation in cellular X-ray microscopy. *J Struct. Biol.* 2012 **178**:29–37.

45 Otón J, Pereiro E, Conesa JJ, Chichón FJ, Luque D, Rodríguez JM, et al. XTEND: extending the depth of field in soft X-ray tomography. *Scientific Reports* 2017 7:45808.

46 LeGros MA, McDermott G, Uchida M, Knoechel CG, Larabell CA High-aperture cryogenic light microscopy. *Journal of Miscroscopy* 2009 **235**:1–8.

47 Smith EA, Cinquin BP, McDermott G, Le Gros MA, Parkinson DY, Tae Kim H, et al. Correlative microscopy methods that maximize specimen fidelity and data completeness, and improve molecular localization capabilities. *Journal of Structural Biology* 2013 **184**:12–20.

48 Cinquin B, Do M, McDermott G, Walters AD, Myllys M, Smith EA, et al. Putting molecules in their place. *Journal of Cellular Biochemistry* 2014 **115**:209–16.

49 Kapishnikov S, Leiserowitz, Yang Y, Cloetens P, Pereiro E, Awamu Ndonglack F, et al. Biochemistry of malaria parasite infected blood cells by X-ray microscopy. *Scientific Reports* 2017 **7**:802.

50 Kapishnikov S, Grolimund D, Schneider G, Pereiro E, McNally JG, Als-Nielsen J, et al. Unraveling heme detoxification in the malaria parasite by X-ray fluorescence microscopy and soft X-ray tomography. *Scientific Reports* 2017.

51 Howells MR, Beetz T, Chapman HN, Cui C, Holton JM, Jacobsen CJ, et al. An assessment of the resolution limitation due to radiation-damage in X-ray diffraction microscopy. *Journal of Electron Spectroscopy and Related Phenomena* 2009 **170**(1–3):4–12.

52 Humbel B. *Freeze-Substitution. Handbook of Cryo-Preparation Methods for Electron Microscopy*, Boca Raton, FL: CRC Press; 2008; 319–41.

53 Schertel A, Snaidero N, Han H-M, Ruhwedel T, Laue M, Grabenbauer M, Möbius W Cryo FIB-SEM: volume imaging of cellular ultrastructure in native frozen specimens. *J Struct Biol* 2013 **184**(2):355–60.

54 Rigort A, Villa E, Bauerlein FJ, Engel BD, and Plitzko JM. Integrative approaches for cellular cryo-electron tomography: correlative imaging and focused ion beam micromachining. *Methods Cell Biol* 2012 **111**:259–81.

55 de Boer P, Hoogenboom JP, and Giepmans BNG. Correlated light and electron microscopy: ultrastructure lights up! *Nature Methods* 2015 **12**(6):503–13.

56 Loussert Fonta C, Leis A, Mathisen C, Bouvier DS, Blanchard W, Volterra A, et al. Analysis of acute brain slices by electron microscopy: A correlative light–electron microscopy workflow based on Tokuyasu cryo-sectioning. *Journal of Structural Biology* 2015 **189**(1):53–61.

57 Schorb, M, Gaechter L, Avinoam O, Sieckmann F, Clarke M, Bebeacua C, et al. New hardware and workflows for semi-automated correlative cryo-fluorescence and cryo-electron microscopy/tomography. *Journal of Structural Biology* 2017 **197**(2):83–93.

58 van der Schaar HM, Melia CE, van Bruggen JA, Strating JR, van Geenen ME, Koster AJ, et al. Illuminating the Sites of Enterovirus Replication in Living Cells by Using a Split-GFP-Tagged Viral Protein. *mSphere* 2016 **1**(4).

59 Liss V, Barlag B, Nietschke M, and Hensel M. Self-labelling enzymes as universal tags for fluorescence microscopy, super-resolution microscopy and electron microscopy. *Scientific Reports* 2015 **5**:17740.

10

Correlative Light- and Liquid-Phase Scanning Transmission Electron Microscopy for Studies of Protein Function in Whole Cells

Niels de Jonge

INM – Leibniz Institute for New Materials and Department of Physics, Saarland University, Saarbrücken, Germany

10.1   Introduction

Cellular function is driven by the molecular machinery of proteins, DNA, and lipids dynamically assembling into macromolecular complexes [1, 2]. An important class of cellular function is communication with the extracellular environment. Receptors in the plasma membranes respond to chemical signals and mechanical conditions by conformational changes, spatial redistribution, and the assembly or reassembly of protein complexes. Information about a certain spatially and temporally localized functional state can be inferred from analyzing the distribution of proteins and their association in protein complexes. Studies of protein complexes, however, are usually conducted by extracting material extracted from many cells, whereby the cells do not remain intact and information about localization is lost. As a result, protein function is incompletely understood [3, 4]. The thus-obtained information about the "average" protein may not necessarily resemble individual proteins in particular because cell function may dramatically differ from cell to cell on account of cell heterogeneity [5]. Moreover, extracting the membrane proteins from the plasma membrane may lead to a manifold of artefacts; for example, certain proteins complexes such as HER2 cannot be extracted without perturbing them [6].

Liquid-phase scanning transmission electron microscopy (liquid STEM) is a new analytical method for studying protein function within the intact plasma membranes of cells in their native liquid environment [7, 8]. The locations of individual membrane proteins are measured with nanometer spatial resolution. Nanoparticle labels are used to tag selected membrane proteins. These targets are then studied with high resolution in an intact cell in liquid, and the experiment is repeated for many cells and under varied conditions. The resolution of liquid STEM allows macromolecular complexes in whole cells in liquid to be studied at the level of *individual subunits*. Liquid STEM is typically combined with light microscopy of cells to navigate cellular regions of interest and to analyze overall protein localization via fluorescence microscopy [9]. It is thus possible to study the stoichiometry of membrane proteins and their distribution in series of selected single mammalian cells and thereby address heterogenic cell populations.

Liquid STEM adds a unique level of analytical characterization possibilities for the study of cellular function.

This chapter includes an overview of the principles of operation and different systems involved. Two examples of research are provided in this chapter showing the potential of this new technique for cell biology. Liquid STEM was used to image individual ORAI11 protein complex subunits labeled with quantum dot nanoparticles [10]. The second example involves the HER2 receptor in breast cancer cells [9]. The chapter discusses the advantages of this technique for cell biological research and finalizes with an outlook pointing into the direction of a new microscopy route in cell biological research.

10.2   Limitations of State-of-the-Art Methods

The interplay of protein complexes within the plasma membrane is in many cases not fully understood, partly because methods are lacking to image them with sufficient resolution within the context of the plasma membrane [3, 4] (see Table 10.1). Biochemical techniques, commonly used to identify protein complexes, use pooled cellular material and do not provide information about the cell of origin of the complexes, nor about the location or environment within a given cell. X-ray crystallography is capable of resolving protein complex structures with atomic resolution but protein crystals have to be made from protein material extracted from many cells. Similarly, single particle cryo-electron microscopy (EM) uses pooled materials, and many thousands of images are then recorded from identical proteins (particles) for averaging to obtain a 3D structure [11]. An example is the 3D structure of the magnesium channel CorA [12]. Direct imaging of proteins in cells is challenging because the involved dimensions are at the nanoscale. Even super-resolution fluorescence techniques [13, 14] lack about an order

Table 10.1 Important analytical techniques used to study the functional state of proteins and their limitations. With permission from John Wiley and Sons [85].

Technique	Limitation
Biochemical methods	Limited to pooled cellular material, proteins do not remain in cells
	Provides information about average responses in a cell population only
Light microscopy	Spatial resolution insufficient to directly view individual protein subunits
	Indirect techniques such as FRET lead to artefacts, e.g. detection of back-to-back neighbors rather than subunits in protein complexes
Flow cytometry	Cells not in adherent state
	Prone to artefacts
Electron microscopy	Samples in vacuum; thus, cells not intact
	Thin cell- or tissue sections needed; challenging to image intact plasma membrane; provides information about few (sections of) cells only
Proximity assay	Does not detect dimers but reflects overall protein proximity, which is heavily influenced by protein concentrations; leads to artefacts

of magnitude in their spatial resolution as needed to resolve the subunits of most membrane protein complexes [15]. A variety of indirect optical techniques exists but all with their specific limitations as discussed in detail elsewhere [9]. For example, Förster resonance energy transfer (FRET) is insufficient as distances in large protein complexes may supersede the FRET distance [16], and fluorescence cross correlation spectroscopy are restricted to very low expression levels of <1 protein complex per square micrometer [17]. In particular, cryo-transmission EM (TEM) is a powerful technique for studying cellular ultrastructure prepared in thin (e.g. 70 nm) cryo sections, and is also combined with fluorescence microscopy [18, 19]. However, the plasma membrane does not remain intact. Protein locations can be determined in membrane patches using freeze fracture [20] but it is practically impossible to obtain stoichiometric protein information combined with localization information within the context of the cell.

10.3 Principle of Liquid STEM

Electron microscopy of liquid specimens became available to a broad scientific community in the past decade on account of new methods based on thin membranes [8, 21–23]. As the technique of liquid cell electron microscopy has become more widespread, it has opened exciting possibilities for solving grand challenges in materials science, chemistry, biology, and other fields [24, 25].

We introduced a novel concept to study proteins in intact eukaryotic cells in their native liquid environment [26–30]. The liquid STEM technology overcomes key limitations in the study of cellular function at the molecular level. Eukaryotic cells in liquid are placed in a microfluidic chamber enclosing the sample in the vacuum of the electron microscope, and are then imaged with STEM (see Figure 10.1a). In order to obtain contrast through water and cell material of several micrometers thickness, nanoparticles (gold nanoparticles or fluorescent quantum dots) are used as specific protein labels [31–34]. Nanometer resolution has been obtained on tagged membrane proteins in whole eukaryotic cells in liquid [8, 28]. The resolution of liquid STEM can be as high as 1.4 nm for 3 μm liquid thickness, as was shown on non-biological test samples [28]. This can be explained on the basis of the atomic number (Z) contrast of STEM [28, 35]. It is also important to emphasize that Brownian motion of the objects under observation did not spoil the resolution[8, 26]. In fact, the movements of floating nanoparticles in close proximity of a SiN membrane were measured to be three orders of magnitude smaller than what would be expected for nanoparticles floating in a bulk liquid [36], and others have even reported atomic resolution in liquid [37].

As an example, to demonstrate the feasibility of the concept, epidermal growth factor receptors (EGFRs) in COS7 cells were labeled with gold nanoparticles. We have developed protocols to achieve specific labeling of EGFR proteins via its ligand EGF using gold nanoparticles of 10 nm diameter [8], and also with QDs providing both a fluorescence signal and Z-contrast for correlative microscopy [32–34, 38]. To avoid label-induced clustering, the labeling was done in two steps, with a formaldehyde fixation step in between [34]. The cells were finally fixed with glutaraldehyde, and imaged as a whole in a liquid layer of ~5 μm thickness [8]. This resolution was achieved well within the limit of radiation damage [8, 39]. Crucial for the study of cell function is the capability to scan many cells and to investigate selected cells with high spatial resolution, which

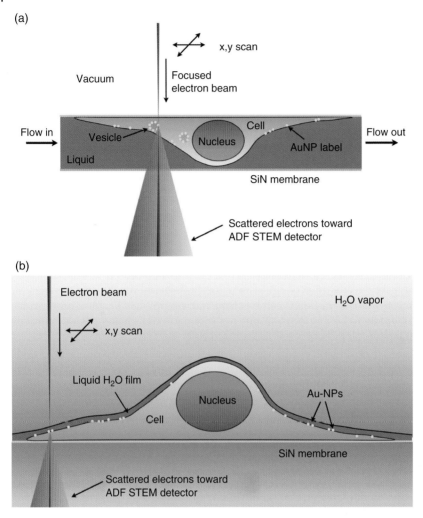

Figure 10.1 Principles of liquid-phase scanning transmission electron microscopy (Liquid STEM). A whole cell is grown on a supporting silicon nitride (SiN) membrane. Proteins labeled with nanoparticles (NPs) reside in the plasma membrane. Imaging is done by scanning a focused electron beam over the cell. Transmitted electrons are recorded with the STEM detector located underneath the sample. (a) The cell is fully enclosed in a microfluidic chamber with two SiN windows. (b) The cell is maintained in a saturated water-vapor atmosphere, while a thin layer of cooled water covers the cell for liquid STEM using environmental scanning electron microscopy (ESEM). With permission from Cambridge University Press [30].

was achieved by combining fluorescence microscopy with liquid STEM and correlating the obtained information [38, 40]. It is also possible to study unfixed cells [30, 40, 41] (e.g. to study nanoparticle uptake) but the question then arises what the influence of the electron beam on the viability of the cells is. We concluded that the electron doses required to achieve nanometer or even tens of nanometer spatial resolution are at least three orders of magnitude over [42] the dose leading to cell death [43]. But future experiments may possibly capture dynamic events of certain biochemical processes of cells. Others have studied, for example, magnetic bacteria in liquid [44] or purified macromolecular protein complexes [45, 46]. Transmission electron microscopy (TEM) most frequently used for biological microscopy would require a much higher dose to obtain a similar resolution for the involved sample thickness but is useful for thinner samples where it obtains nanoscale resolution, e.g. containing viral assemblies [47], or to image organelles in cells at intermediate resolution [48].

It is not always necessary to enclose the cells in the microfluidic chamber. For many studies, it is sufficient to obtain information from the thin outer regions of the cells, and those can be imaged with high resolution using environmental scanning electron microscopy (ESEM)[49, 50] with a STEM detector [7, 51] (Figure 10.1b). We have found this method particularly useful for studies involving many tens of cells [9, 10, 52] because the samples are introduced and imaged with ESEM within a matter of an hour per sample. Full enclosure between two SiN membranes bares the disadvantage that the entire sample should not contain thicker pieces than the desired liquid thickness of typically 5 μm. But a sample with cells often contains a few regions where many cells grow over each other and is thus much thicker. The sample can then not be imaged with high resolution. With ESEM-STEM those regions can simply be avoided and the microscopy be carried out in the thin regions. Yet, ESEM exhibits a lower spatial resolution and handles less thick regions than regular STEM on account of the lower electron energy of 30 keV versus 200 keV typically. Some examples in literature involve an environmental chamber used with regular transmission electron microscopy at 200 keV and beyond [53–55].

A third mode of operation is via sample preparation involving the enclosure of cells under an ultrathin foil formed by graphene sheets [56, 57]. The cells can be enclosed on both sides by sheets or the cells can be supported by a membrane on one side and covered with graphene at the other side [58]. The graphene follows the contours of the cells as a thin, flexible foil. It does not matter if the sample contains rather thick regions. In this case, the sample can be imaged with 200 keV at the highest possible resolution. Dedicated specimen holders are not needed. Disadvantages are that the graphene enclosure is a delicate experimental procedure, and the graphene often shows cracks.

If a study requires a precise localization of proteins but without the need of resolving subunits of protein complexes, while on the other hand superresolution fluorescence techniques are not an option, then an alternative is provided by atmospheric scanning electron microscopy (ASEM) capable of imaging labeled cells in cell an open culture dish [59]. The analysis of intracellular structures and protein distributions has also been demonstrated using ASEM [60]. Also, this method exhibits advantages over conventional methods since the cells are not sectioned. Several other types of liquid enclosures have been developed for SEM including a replaceable capsule [22], and a system with an integrated high-resolution light microscope [61].

10.3.1 Example 1: Determination of ORAI Channel Subunit Stoichiometry by Visualizing Single Molecules Using STEM

The liquid STEM technology is particularly suitable to study the stoichiometry of calcium channels in individual cells. We have developed the protocol to examine the selective Ca^{2+} channel pore-forming ORAI1 protein using liquid phase STEM in ESEM [10]. For this purpose, Jurkat T cells were genetically modified to express ORAI1 with an

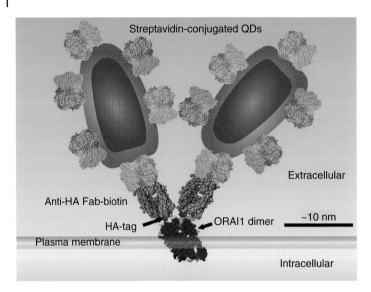

Figure 10.2 Labeling of an ORAI1 dimer (pink) with two quantum dot (QD) probes. An ORAI1 dimer with extracellular positioned hemagglutinin (HA)-tags (red) can bind two anti-HA- fragment antigen-binding fragments (Fabs). The biotinylated Fabs can, in turn, each bind a streptavidin-conjugated QD. Shape and relative size of the shown protein structures are derived from the protein database. With permission from Cambridge University Press [10].

extracellular hemagglutinin (HA)-tag, which is a small nine-amino-acid long tag and – depending on the insertion site – a functionally inert genetic modification [62]. The locations of individual ORAI1 protein subunits were obtained from ESEM-STEM images. Jurkat T-cells expressing HA-tagged ORAI1 were adhered onto the surface of a thin silicon nitride membrane supported by a silicon microchip [31], coated with anti-CD3 and anti-LAF1 antibodies to induce the formation of an immunological synapse [63, 64]. Cells were fixed and labeled with a two-step protocol first with a biotinylated Anti-HA-Fab, binding in a one-to-one stoichiometry to the HA-tag at the second extracellular loop of ORAI1, between transmembrane domains 3 and 4, between amino acids 206 and 207 [65]. The second labeling step consisted of the binding of a Streptavidin quantum-dot (strep-QD) at the biotin moiety. The labeling approach is schematically depicted in Figure 10.2, where the dimensions of the involved structures are drawn approximately to scale. The label is much smaller (less than a third of the size) of an antibody.

Using labels consisting of fluorescent QDs enabled us to examine the cells with light microscopy prior to electron microscopy. The required electron dose was a factor of five below the maximal allowed dose, beyond which radiation damage becomes visible [8, 39]. Typical examples of light microscopic images are shown in Figure 10.3a for direct interference contrast (DIC) light microscopy, showing a topographic view of the cells, and Figure 10.3b, showing the region of the boxed area in Figure 10.3a, displays the red fluorescence emitted by the QDs bound to the ORAI proteins. The distribution of the QD signals at this interface (Figure 10.3b) shows a ring-like pattern whereby most intensity was observed at the rim of the cells. Similar but less prominent ring-like patterns were reported for ORAI1 previously [64], and are also seen for cells with fluorescently-labeled

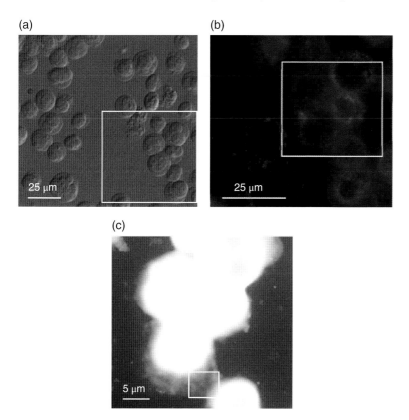

Figure 10.3 Correlative light- and electron microscopy overview images of QD-labeled ORAI1 on Jurkat T cells. (a) Direct interference contrast (DIC) image of Jurkat T cells. (b) Fluorescence image acquired in the box in a. Most cells ORAI1 proteins formed a ring-like pattern consistent with an induced immunological synapse. (c) Liquid-phase ESEM-STEM overview image, displaying the cells marked in b. A rectangle is drawn over a region with thin membrane areas from two neighboring cells for further examination at higher magnification. With permission from Cambridge University Press [10].

actin upon the formation of an immunological synapse [66]. Selected cells were then studied in liquid phase with ESEM-STEM (Figure 10.3c). The QD labels became visible at higher magnification (Figure 10.4a). Enlarged details of this image (Figure 10.4b and c) and of a third cell (Figure 10.4d) reveal the original contrast obtained on the QDs. These images show the spatial distribution and clustering of the labels bound to individual ORAI1 proteins.

For a statistical analysis of the ORAI1 distribution, a total of 4,574 label positions were determined on a total of 818 μm^2 imaged membrane area from 21 examined cells. The collected coordinates were then used to calculate the pair correlation function $g(r)$ according to [9, 67], see Figure 10.5a. The value of $g(r)$ represents the likelihood of two labeled ORAI1 proteins occurring at a certain distance. The function $g(r)$ measures the likelihood of a particle to be found within a certain radial distance with respect to a reference particle [67], whereby $g(r) = 1$ represents a random distribution, and a value >1 indicates clustering with a higher probability than random occurrence. It was found that a distance of $r = 23$ nm occurred much more often than expected

(a)

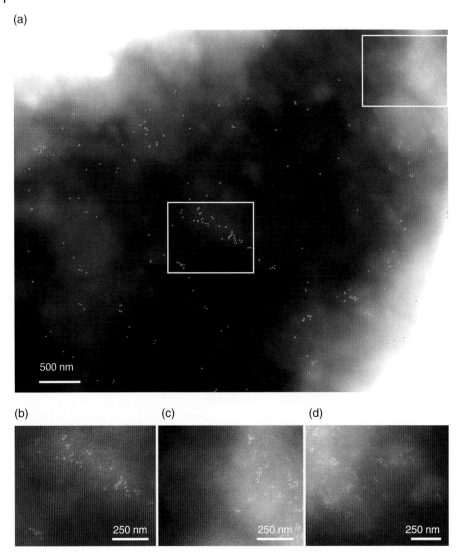

Figure 10.4 ESEM-STEM images of two Jurkat T cells with single QD-labeled ORAI1 proteins. (a) Image recorded from the marked area in Figure 10.3c. Single QD-labels are marked as colored spots in the peripheral plasma membrane regions. Two areas containing larger QD cluster are marked. (b, c) Magnified details from a showing the original contrast obtained on the QD labels. Monomers, pairs, and larger clusters appear in the marked central area (b), and in the upper right area (c) in a. (d) Similar label distributions found in another cell. With permission from Cambridge University Press [10].

based on a random particle distribution. Combined with the molecular model of Figure 10.2, this finding shows that many ORAI1 proteins were adjacent, indicating oligomeric ORAI1 clustering. In addition to this particular distance, the $g(r)$ curve also showed label distances between $\sim 35 - 50$ nm to occur with higher probability than random chance, consistent with the dimensions of QD labels attached to the hexameric state of ORAI1 channels as derived from the crystal structure of the calcium release-activated

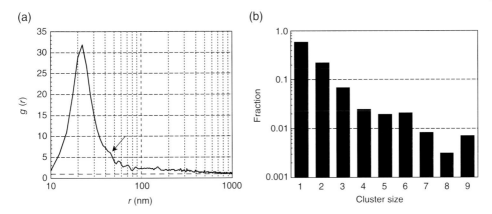

Figure 10.5 Statistical analysis of the label positions. (a) Pair-correlation function (g(r)) versus the label pair distance (r). A sharp peak is visible at 22 nm, and a broad shoulder appears between 35 and 50 nm. The curve slowly approached the random level (dashed line) at 1 μm. (b) The fraction of labels in a cluster of a certain size relative to the total number. From [10].

calcium channel ORAI [68]. The spatial distribution of the labeled ORA1 proteins was further analyzed by applying a cluster-detection algorithm; the results thereof are shown in Figure 10.5. The highest fraction of ORAI1 appeared as monomers, a ~20% of the labels resided as dimers, much larger than what was calculated for a random distribution. Smaller fractions appeared in larger clusters of sizes from 3 to 25 labels. An initial analysis [10] revealed a labeling efficiency of 80%, and we found that dimers and hexamers were present in respective fractions of 0.33 ± 0.06, and 0.06 ± 0.04. Studies of other channels are also readily possible, as we have demonstrated for the calcium-activated chloride channel TMEM16A recently [69].

10.3.1.1 Conclusions

The recombinant Ca^{2+} channel-forming ORAI1 protein with hemagglutinin (HA)-tag was labeled with QDs for correlative fluorescence microscopy and Liquid STEM. A statistical analysis using the pair correlation function showed that many of the proteins reside as pairs, whereas clustering in groups of a size of ~60 nm was also observed. The relative fraction of labels detected as pairs was ~20% and a similar total fraction of ORAI1 was found in clusters most of them assembling three to six labels. These results point toward the possible presence of ORAI1 in multimeric form that may include hexamers, and demonstrate the feasibility of this new method.

10.3.2 Example 2: New Insights into the Role of HER2

To study the distribution of HER2 in the plasma membrane of breast cancer cells at the single molecule level, we have developed a label for HER2 using an affibody peptide [70] (affibody AB, Sweden) conjugated to a QD [9, 34], for which we also filed a patent application [32]. The label contained an affibody attached via a short biotin-streptavidin bond to a QD (see Figure 10.6). Affibodies represent a new type of non-immunoglobulin derived affinity peptide with 10- to 20-fold lower molecular weight than antibodies [71]. One of the main qualities of the affibody is its small dimension of $5 \times 4 \times 3$ nm^3 as determined

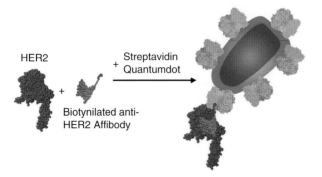

Figure 10.6 Model of the biotinylated anti-HER2 affibody (blue) binding to a single epitope of HER2 (red). The single biotin moiety of the affibody binds a streptavidin (green) conjugated to a bullet-shaped quantum dot (QD). With permission from Cambridge University Press [9].

from the X-ray structure in the protein data bank (pdb entry: 3MZW). The affibody-QD label is bound to HER2 in a 1:1 ratio, on account of a single binding epitope at HER2 for the affibody [70] with a biotinylated C-terminus. The anti-HER2-affibody binding site is at domain III and close to domain IV [70], distant from the dimerization side [72]. The HER2-label binds to a different epitope than trastuzumab [70], so that the presence of the affibody does not alter the effect of trastuzumab, nor does it influence the behavior of HER2 with respect to dimerization or cellular uptake [73, 74]. After exposure to the anti-HER2 affibody, cells are chemically fixed, and the cells are then incubated with streptavidin-conjugated QDs. In this way, labeling-induced clustering [75] is avoided, and a high labeling efficiency is achieved [9].

In contrast to conventional analytic techniques, it is feasible with our method to collect data on receptor membrane expression and stoichiometry in single cells. The usage of liquid-phase STEM presents a paradigm change in the study of the function of membrane proteins in cancer cells [9]. The results shown in Figure 10.7 demonstrate the power of the liquid-phase STEM approach for HER2 cancer cell research. SKBR3 cells, an HER2 overexpressing human breast cancer cell line, were studied with correlative fluorescence microscopy and liquid-phase ESEM-STEM (Figure 10.1b). In contrast to conventional electron microscopy studies, the cells were imaged as a whole and in liquid state, so that the membrane proteins remained in the intact plasma membrane. The locations of individual HER2 receptors were detected using an anti-HER2-affibody in combination with a QD label. Fluorescence microscopy revealed considerable differences of HER2 membrane expression between individual cells, and between different membrane regions of the same cell (Figure 10.7a). Subsequent ESEM of the corresponding cellular regions provided images of individually labeled HER2 receptors (Figure 10.7d). The high spatial resolution of 3 nm, the 1:1 labeling stoichiometry, and the close proximity between the QD and the receptor allowed quantifying the stoichiometry of HER2 complexes, distinguishing between monomers, dimers, and higher-order clusters.

The clustering behavior of HER2 was statistically analyzed via the pair correlation function [67] $g(r)$, calculated from all individual HER2 positions. These positions were automatically detected using a software tool designed by our group for this purpose. In measurements incorporating 14,043 HER2 positions in 11 cells, a sharp peak in the $g(r)$

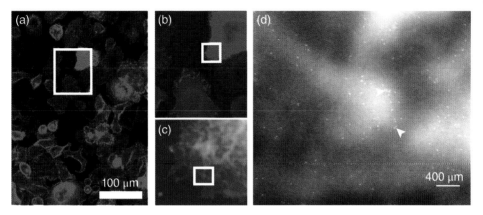

Figure 10.7 Correlative light and electron microscopy overview images of QD labeled HER2 on SKBR3 human breast cancer cells [9]. (a) Fluorescence overview image showing several dozens of cells. Individual cells exhibit a high degree of heterogeneity in their morphology and HER2 membrane expression. (b) Fluorescence image of the cells within the boxed area in a. (c) Liquid-phase STEM image of the boxed region in b recorded at 15,000 × magnification using ESEM. (d) STEM image recorded in the boxed region shown in c at 75,000 × magnification. The locations of individual HER2 receptors labeled with QDs are visible as the bright spots. The brighter background features represent membrane ruffles. Many pairs (homodimers) are visible; two are indicated with the arrowheads. Note that the image looks much different than conventional electron microscopy images showing the cellular ultrastructure. From [9].

function at 20 nm indicated that HER2 was clustered as homodimer (Figure 10.8a). A center-to-center label distance of 20 nm was expected on account of the size of the HER2 dimer, and the two quantum dot labels [9].

HER2 distribution patterns were determined for two distinct cellular regions: membrane ruffles and homogeneous or flat areas. A remarkable difference was found from analyzing $g(r)$ for these different membrane regions. HER2 homodimers (peak at $g(r)$ = 20 nm) appeared in ruffled regions but were entirely absent from homogeneous membrane regions (Figure 10.8b). This observation is a novel scientific finding enabled by the high resolution of liquid-phase STEM combined with its capability to study intact cells. In cancer cells, the highly dynamic membrane ruffles, also referred to as *invadopodia,* are considered to serve as junctions for cellular signaling, and drive motility, invasiveness, and metastasis of cancer cells [76–78]. Our results could thus imply that HER2 homodimers play a role in cancer cell spreading, which is supported by the findings of others showing that HER2 overexpression increased the oncogenic potential in breast epithelial cells [79]

A second imperative finding was the discovery of a small subpopulation of cells with a different phenotype than the average cell [9]. This group of cells was characterized by flat peripheral membrane regions and can possibly be identified as resting (possibly dormant) cells. HER2 homodimers were found to be absent from this subpopulation of cells (Figure 10.8c), even though the concentration of HER2 in the plasma membrane was only ~ 30% lower than in the bulk cancer cells. The absence of HER2 homodimers from these flat cells likely indicates a different intercellular signaling mechanism than the average/bulk SKBR3 cell. In a recent study, we have discovered that this small subpopulation of breast cancer cells responds differently to the prescription drug

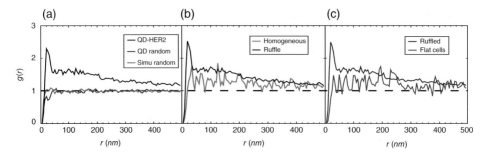

Figure 10.8 Statistical analysis of the spatial distribution of labeled HER2 proteins in 11 SKBR3 cells using the pair correlation function g(r) [9]. (a) g(r) calculated for a total of 14,171 labels exhibited a peak at 20 nm indicating HER2 dimerization. Larger-sized clusters were also observed. The curves of randomly dispersed quantum dots (QDs), and a simulation (simu) of random data, were included as reference. (b) HER2 pairs were absent in cellular areas with a homogeneous or flat membrane topography (3,307 labels), contrasting g(r) in the ruffled areas. (c) HER2 does not appear clustered in the two analyzed flat cells (3,664 labels). Clustering was only observable in cells with membrane ruffles. From [9].

Trastuzumab [80] emphasizing the importance of these single cell/single molecule studies addressing cancer cell heterogeneity.

10.3.2.1 Conclusions

The role of HER proteins is a "hot" topic in cancer research, but despite intensive research efforts using a wide range of techniques over the past decades, this important information about HER2 overexpressing cancer cells without HER2 homodimers was not unveiled before. Our novel scientific findings were obtained as a direct consequence of the high spatial resolution of liquid STEM combined with its capability to study tens of intact cells. The data revealed small subpopulations of breast cancer cells including resting breast cancer cells with different stoichiometry of HER2 and thus with different functional states compared to the bulk cells. HER2 is present in its signaling active form mainly in membrane ruffles, while this form is absent from dormant cells.

10.4 Advantages of Liquid STEM

The liquid STEM technology adds a new level of analytical characterization possibilities, providing a different type of information compared to standard electron microscopy [8, 30]. It exhibits the unique capability of studying protein complexes at the single molecule level in the context of intact cells in liquid. The following key advantages apply for cell biological research:

- The capability to image cells in *liquid* water. It is well known for over a hundred years [81] that the properties of liquid water are vital to life. For example, the structure and properties of water critically influence the secondary and tertiary conformation of proteins, nucleic acids, and membranes. If one changes the water structure, one may also induce changes in macromolecular conformations and cellular structures, including their properties. Apart from a few exceptions, all established electron microscopy methods involve the preparation into solid samples, such as plastic thin sections or cryo-sections. For the case of rapid freezing into amorphous ice, it is frequently assumed that the transition to amorphous ice does not alter the structure of the cell [19]. But it cannot be ruled out that specific effects occur under various experimental conditions. For example, high concentrations of solutes like glycerol are often added to induce vitrification, and cryo-sectioning may introduce various artifacts. Liquid STEM is unique and powerful, because it allows one to visualize cells in liquid water containing normal physiological solutes.

- A spatial resolution better than 3 nm on tagged proteins in whole eukaryotic cells in liquid [7, 8], which is about an order of magnitude higher than achievable with super resolution fluorescence microscopy, while the capability to image whole eukaryotic cells in liquid presents a major breakthrough in electron microscopy.
- Useful for the study of the individual subunits of protein complexes in cells, purified proteins, and virus assemblies in liquid [6,37] [47], and for organelles within intact cells in liquid [40, 44, 48].
- The cells do not need to be genetically modified.
- The high spatial resolution also allows the study of the often high (several tens per square micrometer) endogenous protein concentrations.
- Studies involving more than 100 cells are possible [52].
- Correlative light microscopy and liquid STEM has been firmly established [9, 38, 40]. Spatial correlation of the images is easy because one can use the corner of the silicon nitride window in the microchip as orientation. Moreover, structural changes do not occur because the sample is kept in liquid state.
- Liquid STEM is a rapid method to study cells. The time for sample preparation, imaging, and analysis is comparable to well-established procedures for fluorescence microscopy, but the results have much higher resolution [7–9, 34].
- Last but not least, the costs involved with liquid STEM are moderate for laboratories already using TEM, involving an upgrade with a STEM unit and a fluid specimen holder.

Currently, the main disadvantage is that the technique is not (yet) accepted and adopted within biological community, and only a handful of papers exist applying this technique to study actual biological questions. Liquid-phase electron microcopy has already been accepted in materials science, chemistry, and other fields [24, 25, 82] but until this also happens in biology, the usage of this approach still involves the risks of a pioneering technique. The technology is not yet developed in full and current limitations are:

- The published information obtained on organelles in whole cells typical has a spatial resolution in the tens of nanometer range, but it is possible in principle to obtain better resolution in the thin outer regions of cells, similar to what is achievable with cryo-STEM [83].
- To achieve nanometer resolution on labels, the cells were fixed with glutaraldehyde thus stabilizing the cellular structure. Imaging of labels in unfixed cells has been demonstrated [30] but data about the stability of non-fixed cells in liquid is still lacking.
- The labeling efficiency is not 100% and moreover often unknown. It is then not always clear how to interpret the results. If one observes label positions, the question arises: How closely does it resemble the underlying protein distribution?

- The labels used so far are larger than the proteins, and optimization is needed to reduce the size of the label in the hope that the label will have the least influence on the protein location and the highest possible labeling efficiency is achieved.
- Liquid STEM tomography has not yet been demonstrated but should be possible in principle. A limitation might be presented by the angle of the window in the silicon microchips of 54.7°, so that tilting should occur in the long direction of the silicon nitride window.

10.5 Future Prospects

The concept of liquid STEM presents a paradigm shift in the information that is obtained in biological electron microscopy. Instead of imaging the cellular ultrastructure in thin solid samples, the focus is shifted toward localizing protein labels in whole cells. The images of Figure 10.4 and Figure 10.7 demonstrate a fundamental difference with those of conventional transmission electron microscopy (TEM). Whereas TEM studies usually address the cellular ultrastructure, sometimes with additional immunogold labeling, Liquid STEM only provides low-resolution information of the ultrastructure, which is a direct consequence of the ability to image through the entire cell. Its lower-resolution analog can be found in light microscopy, where fluorescence microscopy is used to study the locations of fluorescent labels. Liquid STEM provides a different type of information because it detects protein positions in intact cells in liquid state and aims at studying many tens of cells, for example, to address cell heterogeneity or to image cells in a series of experiments under varied conditions to study, for example, how protein stoichiometry changes upon external stimuli.

Cryo-STEM can, in principle, be used to detect nanoparticle labels at the edges of whole cells, and this information can be combined with three-dimensional ultrastructure information obtained via STEM tomography [83, 84], but several disadvantages are present. First, the samples would have to be blotted to obtain a sufficiently thin ice layer, and this is a delicate procedure. The cryo-procedures are practically incompatible with experiments requiring the analysis of many cells as needed to find the cellular region of interest or to address cell heterogeneity. Whole cell imaging in combination with the detection of nanoparticle protein labels using cryo-STEM is therefore not used in practice. Once one realizes how simple it is to study cells directly in liquid with correlative light- and electron microscopy, and recognizes the need to study proteins in intact cells and examine larger numbers of cells, the value of this novel technology becomes clear. Instead of providing detailed information about the cellular ultrastructure or the protein structure in a laborious electron microscopic study, liquid STEM focuses on a specific biological process, thereby limiting the amount of information collected per experiment. Similar to fluorescence microscopy, liquid STEM uses labels to tag a specific subset of membrane proteins. The selected targets are then studied with high resolution in an intact cell in liquid, and the experiment is repeated for many cells, and under varied conditions.

A possible option is the study of biochemical processes in unfixed cells. True live-cell electron microscopy is probably impossible as a consequence of the extraordinary large radiation intensity inherent to electron microscopy even at the lowest possible doses of $0.1 \ e^- \mathring{A}^{-2}$ [42], which is already three orders of magnitude above the dose limit for reproductive cell death [43]. Cells do not reproduce and are thus not alive according to the definition of life after exposure to even such low doses [30]. Yet it is possible to study unfixed cells [40, 52], and it would perhaps be possible to record time-lapse image series capturing biochemical processes in cells. A practical approach is to use fluorescence microscopy to pinpoint regions and time points of interest, and then to zoom in with electron microcopy, thereby acquiring a snap shot image of label locations [61]. But careful interpretation of the results would be crucial in order to distinguish between beam-induced effects and physiological processes.

In conclusion, the groundbreaking liquid STEM technology opens up an entirely new field for cellular electron microscopy, providing unprecedented spatial resolution for the study of labeled subunits of macromolecular complexes in eukaryotic cells in their native liquid environment. It is feasible to acquire data of many tens and even hundreds of cells in series of experiments, thereby addressing cell heterogeneity or varying experimental conditions, while still obtaining single-molecule information of endogenous proteins. Correlative fluorescence microscopy and liquid STEM may develop into a key microscopy methodology for future biological research.

Acknowledgments

The author is grateful to D.B. Peckys for co-pioneering liquid STEM and for many discussions, to B. Niemeyer, S. Wiemann, and D.W. Piston for discussions, and to E. Arzt for his support through INM.

References

1 Sali, A., Glaeser, R., Earnest, T., Baumeister, W. (2003) From words to literature in structural proteomics. *Nature*, **422**, 216–225.

2 Baker, M. (2012) Proteomics: The interaction map. *Nature*, **484**, 271–275.

3 Bessman, N.J., Freed, D.M., Lemmon, M.A. (2014) Putting together structures of epidermal growth factor receptors. *Curr Opin Struct Biol*, **29**, 95–101.

4 Valley, C.C., Lidke, K.A., Lidke, D.S. (2014) The Spatiotemporal Organization of ErbB Receptors: Insights from Microscopy. *Cold Spring Harbor Perspectives in Biology*, **6**, a020735-020731-020713.

5 Hanahan, D., Weinberg, R.A. (2000) The hallmarks of cancer. *Cell*, **100**, 57–70.

6 Arkhipov, A., Shan, Y., Das, R., Endres, N.F., Eastwood, M.P., Wemmer, D.E., Kuriyan, J., Shaw, D.E. (2013) Architecture and membrane interactions of the EGF receptor. *Cell*, **152**, 557–569.

7 Peckys, D.B., Baudoin, J.P., Eder, M., Werner, U., de Jonge, N. (2013) Epidermal growth factor receptor subunit locations determined in hydrated cells with environmental scanning electron microscopy. *Scientific reports*, **3**, 2626: 2621–2626.

8 de Jonge, N., Peckys, D.B., Kremers, G.J., Piston, D.W. (2009) Electron microscopy of whole cells in liquid with nanometer resolution. *Proc. Natl. Acad. Sci.*, **106**, 2159–2164.

9 Peckys, D.B., Korf, U., de Jonge, N. (2015) Local variations of HER2 dimerization in breast cancer cells discovered by correlative fluorescence and liquid electron microscopy. *Science Advances*, **1**, e1500165.

10 Peckys, D.B., Alansary, D., Niemeyer, B.A., de Jonge, N. (2016) Visualizing quantum dot labeled ORAI1 proteins in intact cells via correlative light- and electron microscopy. *Microsc Microanal*, **22**, 902–912.

11 Frank, J. (2006) *Three-dimensional electron microscopy of macromolecular assemblies-Visualization of biological molecules in their native state*. Oxford University Press, Oxford.

12 Matthies, D., Dalmas, O., Borgnia, M.J., Dominik, P.K., Merk, A., Rao, P., Reddy, B.G., Islam, S., Bartesaghi, A., Perozo, E., et al. (2016) Cryo-EM Structures of the Magnesium Channel CorA Reveal Symmetry Break upon Gating. *Cell*, **164**, 747–756.

13 Hell, S.W. (2007) Far-field optical nanoscopy. *Science*, **316**, 1153–1158.

14 Lippincott-Schwartz, J., Manley, S. (2009) Putting super-resolution fluorescence microscopy to work. *Nature Methods*, **6**, 21–23.

15 Shivanandan, A., Deschout, H., Scarselli, M., Radenovic, A. (2014) Challenges in quantitative single molecule localization microscopy. *FEBS Lett*, **588**, 3595–3602.

16 Piston, D.W., Kremers, G.J. (2007) Fluorescent protein FRET: the good, the bad and the ugly. *Trends Biochem. Sci.*, **32**, 407–414.

17 Arant, R.J., Ulbrich, M.H. (2014) Deciphering the subunit composition of multimeric proteins by counting photobleaching steps. *ChemPhysChem*, **15**, 600–605.

18 Koning, R.I., Celler, K., Willemse, J., Bos, E., van Wezel, G.P., Koster, A.J. (2014) Correlative cryo-fluorescence light microscopy and cryo-electron tomography of Streptomyces. *Methods Cell Biol*, **124**, 217–239.

19 Kourkoutis, L.F., Plitzko, J.M., Baumeister, W. (2012) Electron Microscopy of Biological Materials at the Nanometer Scale. *Annu. Rev. Mater. Res.*, **42**, 33–58.

20 Cambi, A., Lidke, D.S. (2012) Nanoscale membrane organization: where biochemistry meets advanced microscopy. *ACS Chem. Biol.*, **7**, 139–149.

21 Williamson, M.J., Tromp, R.M., Vereecken, P.M., Hull, R., Ross, F.M. (2003) Dynamic microscopy of nanoscale cluster growth at the solid-liquid interface. *Nature Materials*, **2**, 532–536.

22 Thiberge, S., Nechushtan, A., Sprinzak, D., Gileadi, O., Behar, V., Zik, O., Chowers, Y., Michaeli, S., Schlessinger, J., Moses, E. (2004) Scanning electron microscopy of cells and tissues under fully hydrated conditions. *Proc. Natl. Acad. Sci.*, **101**, 3346–3351.

23 de Jonge, N., Peckys, D.B., Veith, G.M., Mick, S., Pennycook, S.J., Joy, C.S. (2007) Scanning transmission electron microscopy of samples in liquid (liquid STEM). *Microscopy and Microanalysis*, **13(suppl 2)**, 242–243.

24 Ross, F.M. (2015) Opportunities and challenges in liquid cell electron microscopy. *Science*, **350**, aaa9886-9881-9889.

25 de Jonge, N., Ross, F.M. (2011) Electron microscopy of specimens in liquid. *Nature Nanotechnology*, **6**, 695–704.

26 Peckys, D.B., Veith, G.M., Joy, D.C., de Jonge, N. (2009) Nanoscale imaging of whole cells using a liquid enclosure and a scanning transmission electron microscope. *PLoS One*, **4**, e8214.

27 Ring, E.A., de Jonge, N. (2010) Microfluidic system for transmission electron microscopy. *Microscopy and Microanalysis*, **16**, 622–629.

28 de Jonge, N., Poirier-Demers, N., Demers, H., Peckys, D.B., Drouin, D. (2010) Nanometer-resolution electron microscopy through micrometers-thick water layers. *Ultramicroscopy*, **110**, 1114–1119.

29 de Jonge, N., Pfaff, M., Peckys, D.B. (2014) Practical aspects of transmission electron microscopy in liquid. *Advances in Imaging and Electron Physics*, **186**, 1–37.

30 Peckys, D.B., de Jonge, N. (2014) Liquid Scanning Transmission Electron Microscopy: Imaging Protein Complexes in their Native Environment in Whole Eukaryotic Cells. *Microscopy and microanalysis: the official journal of Microscopy Society of America, Microbeam Analysis Society, Microscopical Society of Canada*, **20**, 346–365.

31 Ring, E.A., Peckys, D.B., Dukes, M.J., Baudoin, J.P., de Jonge, N. (2011) Silicon nitride windows for electron microscopy of whole cells. *J. Microsc.*, **243**, 273–283.

32 Peckys, D.B., Dukes, M.J., de Jonge, N. (2014) Correlative fluorescence and electron microscopy of quantum dot labeled proteins on whole cells in liquid. *Methods in molecular biology*, **1117**, 527–540.

33 Peckys, D.B., Bandmann, V., dc Jonge, N. (2014) Correlative fluorescence- and scanning transmission electron microscopy of quantum dot labeled proteins on whole cells in liquid. *Meth. Cell Biol.*, **124**, 305–322.

34 Peckys, D.B., de Jonge, N. (2015) Studying the stoichiometry of epidermal growth factor receptor in intact cells using correlative microscopy. *J. Vis. Exp.*, **103**, e53186.

35 Demers, H., Poirier-Demers, N., Drouin, D., de Jonge, N. (2010) Simulating STEM imaging of nanoparticles in micrometers-thick substrates. *Microsc Microanal*, **16**, 795–804.

36 Ring, E.A., de Jonge, N. (2012) Video-frequency scanning transmission electron microscopy of moving gold nanoparticles in liquid. *Micron*, **43**, 1078–1084.

37 Yuk, J.M., Park, J., Ercius, P., Kim, K., Hellebusch, D.J., Crommie, M.F., Lee, J.Y., Zettl, A., Alivisatos, A.P. (2012) High-resolution EM of colloidal nanocrystal growth using graphene liquid cells. *Science*, **336**, 61–64.

38 Dukes, M.J., Peckys, D.B., de Jonge, N. (2010) Correlative fluorescence microscopy and scanning transmission electron microscopy of quantum-dot-labeled proteins in whole cells in liquid. *ACS Nano*, **4**, 4110–4116.

39 Hermannsdörfer, J., Tinnemann, V., Peckys, D.B., de Jonge, N. (2016) The effect of electron beam irradiation in environmental scanning transmission electron microscopy of whole cells in liquid. *Microsc Microanal*, **20**, 656–665.

40 Peckys, D.B., Mazur, P., Gould, K.L., de Jonge, N. (2011) Fully hydrated yeast cells imaged with electron microscopy. *Biophys. J.*, **100**, 2522–2529.

41 Peckys, D.B., de Jonge, N. (2011) Visualization of gold nanoparticle uptake in living cells with liquid scanning transmission electron microscopy. *Nano Lett*, **11**, 1733–1738.

42 de Jonge, N., Peckys, D.B. (2016) Live Cell Electron Microscopy Is Probably Impossible. *ACS Nano*, **10**, 9061–9063.

43 Reimer, L., Kohl, H. (2008) *Transmission electron microscopy: physics of image formation*. Springer, New York.

44 Woehl, T.J., Kashyap, S., Firlar, E., Perez-Gonzalez, T., Faivre, D., Trubitsyn, D., Bazylinski, D.A., Prozorov, T. (2014) Correlative electron and fluorescence microscopy of magnetotactic bacteria in liquid: toward in vivo imaging. *Scientific reports*, **4**, 6854-6851-6858.

45 Evans, J.E., Jungjohann, K.L., Wong, P.C.K., Chiu, P.L., Dutrow, G.H., Arslan, I., Browning, N.D. (2012) Visualizing macromolecular complexes with in situ liquid scanning transmission electron microscopy. *Micron*, **43**, 1085–1090.

46 Mirsaidov, U.M., Zheng, H., Casana, Y., Matsudaira, P. (2012) Imaging protein structure in water at 2.7 nm resolution by transmission electron microscopy. *Biophysical Journal*, **102**, L15–17.

47 Gilmore, B.L., Showalter, S.P., Dukes, M.J., Tanner, J.R., Demmert, A.C., McDonald, S.M., Kelly, D.F. (2013) Visualizing viral assemblies in a nanoscale biosphere. *Lab on a chip*, **13**, 216–219.

48 Besztejan, S., Keskin, S., Manz, S., Kassier, G., Bucker, R., Venegas-Rojas, D., Trieu, H.K., Rentmeister, A., Miller, R.J. (2017) Visualization of Cellular Components in a Mammalian Cell with Liquid-Cell Transmission Electron Microscopy. *Microsc Microanal*, **23**, 46–55.

49 Kirk, S.E., Skepper, J.N., Donald, A.M. (2009) Application of environmental scanning electron microscopy to determine biological surface structure. *J Microsc*, **233**, 205–224.

50 Stokes, D.L. (2008) *Principles and practice of variable pressure/environmental scanning electron microscopy (VP-SEM)*. Wiley, Chichester, West-Sussex.

51 Bogner, A., Thollet, G., Basset, D., Jouneau, P.H., Gauthier, C. (2005) Wet STEM: A new development in environmental SEM for imaging nano-objects included in a liquid phase. *Ultramicroscopy*, **104**, 290–301.

52 Peckys, D.B., de Jonge, N. (2014) Gold nanoparticle uptake in whole cells in liquid examined by environmental scanning electron microscopy. *Microsc Microanal*, **20**, 189–197.

53 Sugi, H., Akimoto, T., Sutoh, K., Chaen, S., Oishi, N., Suzuki, S. (1997) Dynamic electron microscopy of ATP-induced myosin head movement in living muscle filaments. *Proc. Natl. Acad. Sci.*, **94**, 4378–4392.

54 Matricardi, V.R., Moretz, R.C., Parsons, D.F. (1972) Electron diffraction of wet proteins: catalase. *Science*, **177**, 268–270.

55 Parsons, D.F., Matricardi, V.R., Moretz, R.C., Turner, J.N. (1974) Electron microscopy and diffraction of wet unstained and unfixed biological objects. *Advances in Biological and Medical Physics*, **15**, 161–270.

56 Park, J., Park, H., Ercius, P., Pegoraro, A.F., Xu, C., Kim, J.W., Han, S.H., Weitz, D.A. (2015) Direct Observation of Wet Biological Samples by Graphene Liquid Cell Transmission Electron Microscopy. *Nano Lett*, **15**, 4737–4744.

57 Wojcik, M., Hauser, M., Li, W., Moon, S., Xu, K. (2015) Graphene-enabled electron microscopy and correlated super-resolution microscopy of wet cells. *Nat Commun*, **6**, 7384:7381–7386.

58 Dahmke, N.D., Hermannsdoerfer, J., Weatherup, R.S., Hofmann, S., Peckys, D.B., de Jonge, N. (2016) Electron microscopy of single cells in liquid for stoichiometric analysis of transmembrane proteins *Microsc Microanal*, **22**(S5), 74–75.

59 Nishiyama, H., Suga, M., Ogura, T., Maruyama, Y., Koizumi, M., Mio, K., Kitamura, S., Sato, C. (2010) Atmospheric scanning electron microscope observes cells and tissues in open medium through silicon nitride film. *J Struct Biol*, **169**, 438–449.

60 Maruyama, Y., Ebihara, T., Nishiyama, H., Suga, M., Sato, C. (2012) Immuno EM-OM correlative microscopy in solution by atmospheric scanning electron microscopy (ASEM). *J Struct Biol*, **180**, 259–270.

61 Liv, N., van Oosten Slingeland, D.S., Baudoin, J.P., Kruit, P., Piston, D.W., Hoogenboom, J.P. (2016) Electron Microscopy of Living Cells During in Situ Fluorescence Microscopy. *ACS Nano*, **10**, 265–273.

62 Field, J., Nikawa, J., Broek, D., MacDonald, B., Rodgers, L., Wilson, I.A., Lerner, R.A., Wigler, M. (1988) Purification of a RAS-responsive adenylyl cyclase complex from Saccharomyces cerevisiae by use of an epitope addition method. *Mol Cell Biol*, **8**, 2159–2165.

63 Quintana, A., Pasche, M., Junker, C., Al-Ansary, D., Rieger, H., Kummerow, C., Nunez, L., Villalobos, C., Meraner, P., Becherer, U., et al. (2011) Calcium microdomains at the immunological synapse: how ORAI channels, mitochondria and calcium pumps generate local calcium signals for efficient T-cell activation. *EMBO J*, **30**, 3895–3912.

64 Alansary, D., Bogeski, I., Niemeyer, B.A. (2015) Facilitation of Orai3 targeting and store-operated function by Orai1. *Biochim Biophys Acta*, **1853**, 1541–1550.

65 Gwack, Y., Srikanth, S., Feske, S., Cruz-Guilloty, F., Oh-hora, M., Neems, D.S., Hogan, P.G., Rao, A. (2007) Biochemical and functional characterization of Orai proteins. *J Biol Chem*, **282**, 16232–16243.

66 Nolz, J.C., Gomez, T.S., Zhu, P., Li, S., Medeiros, R.B., Shimizu, Y., Burkhardt, J.K., Freedman, B.D., Billadeau, D.D. (2006) The WAVE2 complex regulates actin cytoskeletal reorganization and CRAC-mediated calcium entry during T cell activation. *Curr Biol*, **16**, 24–34.

67 Stoyan, D., Stoyan, H. (1996) Estimating pair correlation functions of planar cluster processes. *Biom. J.*, **38**, 259–271.

68 Hou, X., Pedi, L., Diver, M.M., Long, S.B. (2012) Crystal structure of the calcium release-activated calcium channel Orai. *Science*, **338**, 1308–1313.

69 Peckys, D.B., Stoerger, C., Latta, L., Wissenbach, U., Flockerzi, V., de Jonge, N. (2017) The stoichiometry of the TMEM16A ion channel determined in intact plasma membranes of COS-7 cells using liquid-phase electron microscopy. *J Struct Biol*, **199**, 102–113.

70 Eigenbrot, C., Ultsch, M., Dubnovitsky, A., Abrahmsen, L., Hard, T. (2010) Structural basis for high-affinity HER2 receptor binding by an engineered protein. *Proc Natl Acad Sci U S A*, **107**, 15039–15044.

71 Lofblom, J., Feldwisch, J., Tolmachev, V., Carlsson, J., Stahl, S., Frejd, F.Y. (2010) Affibody molecules: Engineered proteins for therapeutic, diagnostic and biotechnological applications. *FEBS Letters*, **584**, 2670–2680.

72 Arkhipov, A., Shan, Y.B., Kim, E.T., Dror, R.O., Shaw, D.E. (2013) Her2 activation mechanism reflects evolutionary preservation of asymmetric ectodomain dimers in the human EGFR family. *Elife*, **2**, e00708.

73 Orlova, A., Nilsson, F.Y., Wikman, M., Widstrom, C., Stahl, S., Carlsson, J., Tolmachev, V. (2006) Comparative in vivo evaluation of technetium and iodine labels on an anti-HER2 affibody for single-photon imaging of HER2 expression in tumors. *J Nucl Med*, **47**, 512–519.

74 Steffen, A.C., Wikman, M., Tolmachev, V., Adams, G.P., Nilsson, F.Y., Stahl, S., Carlsson, J. (2005) In vitro characterization of a bivalent anti-HER-2 affibody with potential for radionuclide-based diagnostics. *Cancer Biother Radiopharm*, **20**, 239–248.

75 Brown, E., Verkade, P. (2010) The use of markers for correlative light electron microscopy. *Protoplasma*, **244**, 91–97.

76 Weaver, A.M. (2006) Invadopodia: specialized cell structures for cancer invasion. *Clin Exp Metastasis*, **23**, 97–105.

77 Feldner, J.C., Brandt, B.H. (2002) Cancer cell motility – on the road from c-erbB-2 receptor steered signaling to actin reorganization. *Exp Cell Res*, **272**, 93–108.

78 Brix, D., Clemmensen, K., Kallunki, T. (2014) When Good Turns Bad: Regulation of Invasion and Metastasis by ErbB2 Receptor Tyrosine Kinase. *Cells*, **3**, 53–78.

79 Ingthorsson, S., Andersen, K., Hilmarsdottir, B., Maelandsmo, G.M., Magnusson, M.K., Gudjonsson, T. (2015) HER2 induced EMT and tumorigenicity in breast epithelial progenitor cells is inhibited by coexpression of EGFR. *Oncogene*.

80 Peckys, D.B., Korf, U., Wiemann, S., de Jonge, N. (2017) Liquid-phase electron microscopy of molecular drug response in breast cancer cells reveals irresponsive cell subpopulations related to lack of HER2 homodimers. *Mol Biol Cell.*

81 Henderson, L.J. (1913) *The fitness of the environment*, New York.

82 Ross, F.M., Wang, C.M., de Jonge, N. (2016) Transmission electron microscopy of specimens and processes in liquids. *Mrs Bulletin*, **41**, 791–799.

83 Wolf, S.G., Houben, L., Elbaum, M. (2014) Cryo-scanning transmission electron tomography of vitrified cells. *Nat Methods*, **11**, 423–428.

84 Hohmann-Marriott, M.F., Sousa, A.A., Azari, A.A., Glushakova, S., Zhang, G., Zimmerberg, J., Leapman, R.D. (2009) Nanoscale 3D cellular imaging by axial scanning transmission electron tomography. *Nat Methods*, **6**, 729–731.

85 de Jonge, N. (2018) Correlative light microscopy and liquid-phase scanning transmission electron microscopy of labelled membrane proteins in whole mammalian cells. *J. Microsc.*, **269**, 134–142.

11

Correlating Data from Imaging Modalities

Perrine Paul-Gilloteaux[1] and Martin Schorb[2]

[1] *Structure Fédérative de Recherche François Bonamy, CNRS, INSERM, Université de Nantes, France*
[2] *European Molecular Biology Lab (EMBL), Heidelberg, Germany*

11.1 Introduction

One of the key tasks in correlative light and electron microscopy (CLEM) is the registration of multiple independent coordinate systems in either two or three dimensions, leading to the spatial correlation of data. The process of overlaying images (two or more) of the same scene taken at different times, from different viewpoints, and/or by different sensors is referred to as image registration. There are many scientific applications that require image registration such as the alignment of different fluorescence channels or of images in a time-lapse or serial-sectioning experiment. For example, images of a time lapse are aligned to remove image shifts caused by thermal drift or mechanical instability of the microscope stage. The registration will seek for a global alignment, not taking into account any local movement such as cell movement. Those images will always share a multitude of common features that the alignment procedure can utilize as the acquisition modality is the same for all images. In CLEM the difficulty is, that while we observe the same sample, it is imaged by extremely different imaging modalities – light microscopy (LM) and electron microscopy (EM). The intrinsic difference in image formation between these two methods is what gives CLEM its strength, the complementary localization information of specific dynamic processes (from FM) in a global morphological context (from EM). As the visual appearance of image features in general differs a lot in the two datasets (Figure 11.1), the matching of landmark features poses a challenging image analysis problem.

When speaking about image registration in the computer vision field, usually two main approaches can be differentiated: intensity-based and feature-based. In the intensity-based approach, one image is successively transformed until it reaches a good similarity in intensity with the target image (sometimes after some intensity transformation so it looks more similar to the other one). This similarity measurement is called metric (e.g., correlation, mutual information metrics). In the feature-based approach, only specific parts of the two images, with common information in both images, are extracted and expressed as a simplified representation of a coordinate frame. Then these features are

Correlative Imaging: Focusing on the Future, First Edition. Edited by Paul Verkade and Lucy Collinson.
© 2020 John Wiley & Sons Ltd. Published 2020 by John Wiley & Sons Ltd.

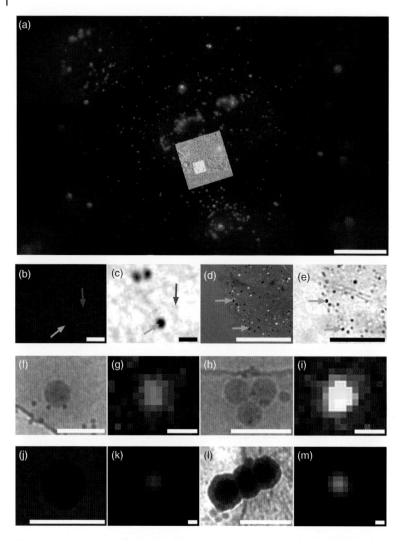

Figure 11.1 Diversity of features and difficulties of multiscale features: Example of data sets of the same specimens imaged in electron and fluorescence microscopy. (a) Typical difference in scale between LM (red), low magnification EM (gray) and high magnification EM (yellow) images. Scale bar 10 microns; (b–c): High magnification EM and corresponding bright field LM image showing a mixture of melanosomes (natural features, example green arrow) and beads in red in C (artificial features, example red arrow). Scale bar 1 micron. (d–e) High magnification EM and bright field LM showing melanosomes (natural features, example green arrows) Scale bars: 10 microns. (f–g) High \ micron. (h–i) As in (f–g), but with three beads Scale bars: 0.5 microns. (j–k) Q-dot in high magnification EM and fluorescence image. Scale bars: 0.1 micron. (l–m) As in j–k, but with 3 quantum dots Scale bars: 0.1 micron. Data acquired by Xavier Heiligenstein (a–d and j–m) and Martin Schorb [1] (f–i).

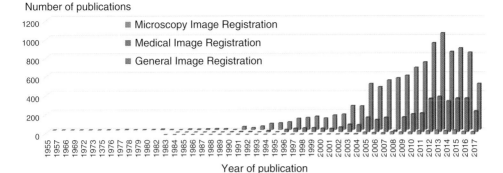

Number of publications

Figure 11.2 Number of publications per year presenting methods of general image registration vs medical image registration and microscopy image registration toward 2017, as given by a PubMed research on the following keywords: *image registration for general image registration, microscopy image registration,* and *medical image registration.* The proportion of publications associated with *correlative microscopy image registration* was 10 in total and does not appear on the graph.

successively transformed until they match the coordinates of the target features. The selected transform is then applied to the whole image. To unite the information of the two modalities EM and LM, one of the most efficient current approaches is to add artificial landmarks that serve as fiducial markers in LM and EM, or to use natural common visible features as landmarks, meaning using the feature-based approach. But also intensity-based approaches are of interest for CLEM applications, in particular when intermediate images at each stage can be acquired.

In this chapter, we want to present our view on past and current research paving the way for the future of image registration in CLEM. We would additionally like to introduce initiatives not directly related to CLEM, but which we expect to potentially have a big impact in the field if they can be adapted to the specific needs of CLEM. While registration procedures in general are a very active research field, there is only little development in biology compared to, e.g., the medical imaging community. Also, its role in CLEM is not a large factor when it comes to publications (Figure 11.2).

The difference in scale and field of view between LM and EM make developing CLEM workflows a real team game. Limitations and challenges cannot be overcome solely with image registration itself but also need to be addressed by rethinking labeling or sample preparation strategies. Image registration is one key player in CLEM for which no generic solution across different types of applications exists so far. Nevertheless, we envision that developments in the field could drastically reduce the time spent on it during a typical CLEM experiment.

First, we present the different stages of current CLEM workflows where image registration comes into play. We will then quickly explain or remind the reader of the key concepts and elements involved in image and volume registration. For an overview of the different approaches for registration, the interested reader is also referred to this review [2].

Finally, we will explore how some actual development could drastically change the field. Our goal is not to be exhaustive, but rather to give some general views, illustrate them by examples, and maybe trigger more developments in the field of CLEM by the bio image informatics community.

11.2   Registration during CLEM Stages

Image registration happens at various steps in a CLEM workflow (Figure 11.3). For the rest of this chapter, we assume that the regions of interest (ROIs) have been identified in light microscopy (LM). The region of interest can be a cell, or a particular cellular compartment, or even a tiny vesicle inside or outside a cell. The coordinates of this ROI, i.e., its position in the LM image, can be used for guiding and targeting the specimen preparation procedure, especially the ultramicrotomy (section: Registration to guide the sample preparation). Another step of registration is necessary to guide the acquisition at the second microscope (typically the EM) and adjust the field of view of the acquiring microscope to this feature (section: registration to guide the acquisition). In a last, and most accurate registration step, the acquired data from both modalities is registered after finishing all imaging (section: Post-acquisition registration (accurate relocation)).

11.2.1   Registration to Guide Sample Preparation

The vast majority of EM techniques require a step to physically reduce the volume and/ or thickness of the specimen in order to prepare it for EM imaging. The localization of the ROI can already be used to target the necessary manipulation. Various approaches

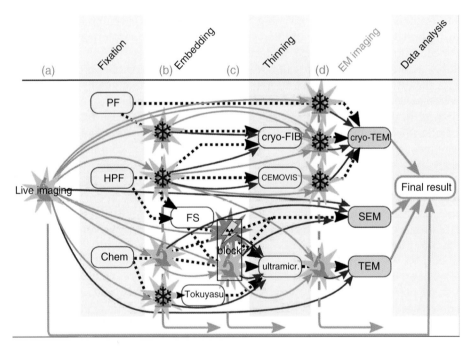

Figure 11.3 Schematic depiction of the possible registration steps during CLEM workflows. The symbols indicate the optional light microscopy steps during the process. These can happen (a): before fixation imaging the living specimen, (b): after fixation, (c): after embedding (micro-CT indicated in blue), or (d): right before EM imaging. Dark-blue dotted arrows indicate typical workflows, orange arrows indicate image registration between different imaging steps, and red arrows indicate coordinate information that used for targeting during a later stage of the process. Image information from all steps is used throughout the literature to register and compare it with the EM image data.

exist to guide the ultramicrotomy or FIB milling either in the room temperature case [3–5] or when preparing cryogenic specimens for EM [6–8], but they mainly all rely on the visual identification of features creating a coordinate system. Features can be a scratch on the sample, or the shape of the sample itself. The purpose is to be able to identify a few reference points that will be used to roughly locate the region of interest (5 mm below the scratch center for example). Fluorescent microscopes have been added to ultramicrotomes to ease the guidance, paving the way to facilitate the automation of this step [9]. It then allows the use of more intrinsic features to find back the position of interest, by using for example some specific patterns in nuclei positions.

11.2.2 Registration to Guide the Acquisition

11.2.2.1 Software Packages

Several software packages that control an electron microscope and its detectors offer the possibility to use image data from external sources (LM) to target scanning electron microscopy (SEM) acquisition. The idea that both Zeiss – with the ZEN software and the "Shuttle and Find" module [3] – and FEI / Thermo Fisher – with MAPS [10] – pursue is that the same acquisition software runs on both the light and the electron microscope. Special landmark features on the specimen holder allow the registration of two modalities based on physical stage coordinates. In that case, the user has to select two or three points of interest corresponding to a specific mark on the sample holder that will be manually or automatically found in the EM image using robust descriptors [11, 12], for a global accuracy of less than 8 microns. The physical coordinates of the area of interest are transferred from LM to EM, in a coordinate system that is completely attached to the sample (the sample holder) and microscope stage.

There is also MosaicPlanner [13], an open source acquisition software package based on MicroManager [14] that can be used to register FM and SEM image stacks based on intrinsic image features selected manually by the user. In MosaicPlanner, two points are selected and the registration is refined using an intensity-based rigid registration (here using normalized correlation as metric).

When using an integrated light and scanning electron microscope (SEM) the photonic signal generated inside the specimen by a focused electron beam can be used to accurately align the two imaging systems [9].

Software packages for transmission electron microscopy (TEM) acquisition, such as the JEOL "pop" software or the open source package SerialEM [15] that can operate on a number of microscopes from different manufacturers offer the possibility to import data from different imaging modalities and then manually registering to the current EM coordinate system. Rather recent developments offer new, flexible ways of importing coordinates from any external source into SerialEM.

11.2.2.2 Typical Features and Fields of View

The manual search for ROIs, which could be as small as a few microns, is almost always excessively time consuming or sometimes even impossible due to the amount of potential target features. Bright-field imaging that visualizes cellular morphology in a similar way as low-magnification EM does, is commonly used for guiding the registration when using content-based approach.

With specifically patterned grids (e.g., commercial finder grids) and global feature-recognition, a registration of image maps that cover a large area of the specimen would be possible. The outline and location of the section(s) on the specimen support can also be a promising feature. From experience gathered in our own experiments using manually assigned landmark points [16], the possible accuracy of a registration on the global TEM grid is severely limited (to approx. 5–10 µm) by both the necessity to stitch tiled images to cover a sufficient field of view (FOV) and the inherent optical distortions when operating a TEM at very low magnifications. Automation at this level will only help to coarsely identify the regions where to acquire. Combining it with a motorized holder that enables the rotation of the specimen while inside the EM (such as the Fischione Model 2045 Motorized Dual-Axis Tomography Holder) could, however, considerably simplify the necessary follow-up registration at higher EM magnifications.

For the registration on EM acquisition, the required accuracy is inherently defined by the desired pixel size and thus the available FOV. The bigger the FOV is, the smaller the resolution, but the easier the registration will be. For this reason, the acquisition in EM is usually done at low magnification, followed by higher magnification once the region of interest is found back.

Other studies have made use of additional imaging steps, allowing to relieve the problem of the limited field of view offer by EM. This is in particular the case for the study where Karreman and colleagues [4, 17, 18] used an additional micro CT-scan on resin block in order to accurately localize a cell of interest in a the coordinate system defined by blood vessels visible both in LM and in the micro CT. A similar approach is taken when using an integrated light microscopy observation together with the EM, but with the critical prerequisite of retaining fluorescence in the resin.

11.2.3   Post-Acquisition Registration (Accurate Relocation)

All the steps described above result in the acquisition of the same area of the sample, but the pixel matching is usually far from accurate at this stage. An additional step of registration is then performed in all CLEM workflows after acquiring all image data. This registration is mostly based on the image content itself or on accurately localized features within such as microspheres.

11.2.3.1   Software and Approaches for Post-Acquisition Registration

Publications using CLEM have used a wide range of methods for this last step of CLEM registration, starting from entirely manual alignment, using, e.g., Photoshop and user visual feedback, to semi- and even fully automatized methods. They can be based on artificial landmarks (e.g., beads), or on natural ones (e.g., mitochondrial pattern [19]).

Semiautomated approaches generally use manual selection of landmarks, successively paired on the two images, using opensource software platforms such as Icy (with EC-CLEM [20]), Fiji (with TurboReg [21]), and more recently BigWarp [22]), Scipy [23], or standalone software such as 3D Correlation toolbox [6]. Commercial platforms are also used, such as MATLAB (with cpselect and imtransform [24]) or AMIRA (FEI / Thermo Fisher) with the LandmarkWarp function [5, 25]. Usually it starts with identifying natural features (e.g., cells contours, nuclei pattern) and is refined by fiducial beads or small natural landmarks, such as melanosomes.

Some of these methods proposed also use an automatic refinement of the localization of landmarks, based on a Gaussian fitting of microsphere signals for example [1, 26], or based on an intensity based registration after this first accurate initialization [21, 27], using one of the similarity metrics described above. Note that other software tools could allow this two-step process such as Matlab with imregister, or the ITK library, widely used in the medical field. An example of software implementing this library is Elastix [28].

Some fully automated methods (so-called automated if ignoring the step of tuning correctly the parameters of these methods, which should be done once for one data type but could then be theoretically applied to a set of similar images) have also been proposed specifically for CLEM. Some are based on the automatic detection of features and their automatic matching [2, 20, 29, 30], where the specificities of CLEM discrepancy of presence of features can be taken into account by appropriate optimization such as RANSAC [20, 31]. Indeed, some features appearing in EM may not be fluorescent in LM, or not detected correctly in one of the modalities. This approach is, for example, made available in Icy through EC-CLEM AutoFinder. Other approaches also use richer features than point localization, such as patterns defined by these localization, for example constellations as used for selective plane illumination microscopy reconstruction, available on ICY [31] but these have not been applied or adapted to CLEM yet.

It has to be underlined that while the human vision adapts itself very easily to the difference of appearance of common features in two modalities, computer vision approaches still need to adequately tune to a particular CLEM workflow. The difficulty with the existing similarity metrics (e.g., proposed in Amira, Matlab or Elastix) is that they are not sensitive enough to deal with the huge discrepancy of appearance, and in particular the amount of features in EM not visible in LM. This may work for well identifiable structure in EM such as nuclei or mitochondria, assuming that they have been segmented on the EM image. Then the registration between mitochondria segmented pattern and a fluorescent channel only showing a probe for mitochondria, such as MitoTracker, should be quite direct [19]. They are then only used for now in the very last step of the registration as stated above. However some papers have proposed ways to overcome this limitation, using the same philosophy: transforming the intensity values of one or both modalities to come back to a monomodal registration problem, easier to deal with than a multimodal one. This is the case, for example, in [32], where both LM and EM images are transformed using a Laplacian of Gaussian filtering, with an automatic feature scale detection, followed by a robust optimization of an intensity based metric. Another example was published in [33], where the LM image was transformed to EM image based on a machine-learning approach: patches of preregistered images were used to create a catalog of image analogies between the two modalities, allowing predictions for the rough appearance in EM of a LM dataset.

Most of the listed software propose different basis of transformation, rigid or nonrigid, and a majority of them can work on 3D data sets, more or less conveniently and with more or less data preprocessing. Only a few softwares allow mixed registration (i.e., 2D to 3D registration) to our knowledge: EC-CLEM (both in manual and automatic mode with AutoFinder [30]) and 3D Correlation Toolbox [6].

In all these steps, processing big 3D images is still a challenge, already addressed (e.g., for manual pairing of point and nonrigid transformation using BigWarp in Fiji [22]). This software takes advantage of the conversion of the image data to a specific file format, similar to a database of pixels, and of a powerful library to read and write these huge files. We envision that this kind of data manipulation and new format will become the standard in processing data, but also to visualize the result of registration.

11.2.4 Trust in Alignment: Accuracy in Practice

For being as accurate as possible in matching structures of interest, one important question is to evaluate the correlation of two structures indicated by the spatial registration. The difference in resolution between the two modes of imaging complicates the task. An error of a single pixel in LM can correspond to tens or hundreds of pixels in the EM image. In addition to the resolution limitation in LM, most of the automated approaches described above use only a limited number of points.

Computing the accuracy of the registration is one of the central problems of CLEM, in particular because the structure of interest is usually not marked in both modalities. Thus, it is essential to confirm the correlation is correct and unbiased.

The gold standard is to use artificial fiducial markers with a high localization accuracy. The registration error is then evaluated by measuring the remaining discrepancy in position extracted from LM and from EM [1, 26].

In previous publications [1, 24, 34], an iterative leave-one-out method was used, estimating the registration error from the average error of localization of beads not used for the registration and was therefore empirical. Other recent work tried to find a theoretical estimation of the error, using the Cramer Rao limits [35]. In our previous work [20], we used a paradigm validated in the medical field for 2D and 3D rigid registration [36, 37]. This was originally developed for image guided surgery, but can also be applied to CLEM. Interestingly, the average expected accuracy of a rigid registration can be estimated without the need of accurately localized fiducials.

However, bimodal probes as landmarks seem to be without alternative. A sufficiently dense set of fiducials allows to estimate nonrigid distortions and can be seen as the only valid metrology tool for CLEM up to now [24, 38]. Immunogold labeling of fluorescent molecules can be used to confirm the correct localization in EM, with the drawback for some discrepancies in position [39].

11.3 Registration Paradigm

The central paradigm in registration (i.e. spatial alignment) is usually represented as follows [40–43].

The purpose is to find the transformation that minimizes the differences between common elements in two images – a "source image" and a "target image" – and then apply this transformation to identify and characterize the features of interest. The source image is the image which will be transformed to overlay the target image, which will not be modified. In the CLEM case, we are in a typical multimodal registration framework, i.e. combining images from different modalities (LM and EM). In contrast, an example of a monomodal registration would be the alignment and combination of 2D images from serially cut sections resulting in a reconstructed 3D stack. While developing or selecting a registration method, the following elements should be identified, and will allow to differentiate different software and algorithms.

11.3.1 Image Features to Guide the Registration

These can be direct visual image characteristics, such as the intensity of voxels/pixels, or features such as points, lines, or other geometrical elements extracted from the images. It can also be more complex features such as vectors between groups of points or descriptors extracted from the image and presenting interesting properties, such as scale invariant descriptors (SIFT) or rotation invariant descriptors. Features can also be obtained from derived data such as variance maps or gradient flows that the image contains but that are not directly visible. What is important to note here is that errors can occur both during identifying the features, as well as accurately localizing them in both image modalities. This error in localization will then lead to an error in registration (called fiducial localization error in Section 11.4).

In most of the approaches now used for CLEM, these features are the location of artificial or natural landmarks [20, 29], or the intensity values of the image itself. In order to increase similarity between the two images, some filters may be applied [32, 33]. Although we want to have clear features for registration available, we do not like to interfere with the observation of the feature of interest especially during FM. For example for intensity-based registration, a specific channel of fluorescence that does not contain any signal from the feature of interest should be used for the registration with EM. Bright field images of nuclei could be used for the registration, while a fluorescence image contains the experimental information. The signal of beads used for the registration with EM can be imaged in a separate fluorescence channel. This additional channel(s) can also serve to distinguish landmarks whose signal is visible in the channel of interest from the target signal [1, 34]. This would avoid covering the signal of interest by one of the landmarks. In addition, this avoids biasing the registration using assumption of the correspondence of region of interest.

11.3.2 Distance Function

This is the mathematical function giving a score of similarity between features or pixel intensities. For example, for point features it can be simply the difference of position in point (mean square distance), or for intensity, a metric called mutual information or also intensity normalized correlation, both well adapted metrics for multimodal registration. The distance function, or also called energy function or cost function, can also be modified to implement robust approach, i.e., deciding to weight the matching of features as a function of their quality in order to down-weight the importance of points that do not follow the behavior of the majority, and thus are likely outliers.

11.3.3 Transformation Basis

One of the key elements of registration is the choice of which family of transformation to choose to align the images and match common elements between LM (source image) and EM (target image) (Figure 11.4). The goal of the registration procedure is to find the ideal values of the parameters of the proposed transformation basis. The choice of the transformation basis means deciding for either a shape preserving transformation (e.g. rotation, translation, isotropic scaling, or flipping) or a transformation that will modify the shapes (such as anisotropic scaling or curved transformations) (Figure 11.4a). For each type of transformation several mathematical tools and representations exist, that will lead to different results in the alignment. For example, a rigid transformation (shape-preserving) can be represented by rotations and translation along main x-, y-, and z-axes. These transformations can be mathematically phrased using matrix or quaternion notation. Curved transformations (also called nonrigid or elastic) require particular attention. Indeed, a potential infinite number of bases exist to represent this

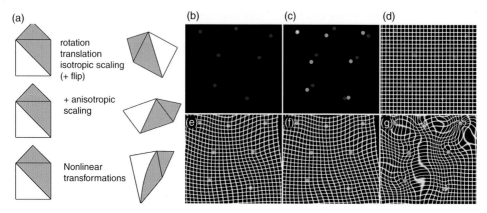

Figure 11.4 Choice of transformation: (a) Transformation basis families; (b-g) Illustration of different bases from one particular family of curved deformation ("B-Spline") allowing several degrees of freedom. Transformations were generated using the software package Elastix [28]. Correlation is used as the metric; a pyramidal approach is used. (b) source image shown (moving image). (c) Target image in green (fixed image) on which the source image is superimposed in red. (d) Reference grid, which will undergo the same transformation as the source image (superimposed in red) to indicate the deformation applied to other pixels in the image. (e-g) shows grids undergoing the transformation leading to a perfect matching of the source points to target (in green), E using a B-spline of order 2 and a support grid of 16 pixels, F with a B-spline of order 3 and a support grid of 16 pixels, and G with a B-spline of order 3 and a support grid of 1 pixel.

family of transformations (e.g. polynomial basis, thin plate spline, B-Spline, daemons) and while they generally all allow a perfect match of extracted features, what happens to the other pixels in the image under the transformation can be drastically different (Figure 11.4b). We must also note that having a perfect match of extracted features may not be the ideal solution for the features of interest: for example in the case of points localized with a certain error of localization, inherent to the physical observation, matching perfectly to another imperfectly localized point will not lead to an optimal transformation. Another aspect of the choice of transformation basis is whether the transformation is global (one basis for the full image/volume) or local (i.e. different parameters for different blocks on the image).

11.3.4   Optimization Strategy

While some direct techniques based on solving matrices equation can be used to compute an optimal transform, called closed-form methods, most of the registration methods are iterative. Determining the best transformation by minimization of the distance metrics between intensity or extracted features is called the *optimization strategy*. Without an optimization strategy, all possible values of parameters would need to be tested (leading to infinite combinations). This approach is called exhaustive search and is never used in practice. Optimization methods can help to decide for a clever choice of the set of parameters for which the distance should be computed next, and is a field of research by itself in computer science. Some well-established approaches are the gradient descent, simulated annealing, or the RANSAC Procedure (RANdom SAmple

Consensus). In addition, the optimization is usually done at several levels of resolution of the images, such as using a pyramidal approach, where parameters are first computed at a coarse level on down-sampled versions of the images, and then refined until reaching the original resolution of the images.

11.4 Envisioned Future Developments

Based on the current advances in the field of CLEM, we envision the main developments to happen in the hybrid use of integrative microscopy and correlative microscopy, but also on the registration algorithms themselves based on a better incorporation of a priori knowledge on the samples and imaging processed. In addition, we discuss the possibilities of using artificial intelligence to enhance the field, and the interaction with data using augmented reality.

11.4.1 Integrative Microscopy versus Correlative Microscopy

When fluorescence can be retained in the specimen imaged by EM, integrated light microscopes allow for a direct registration of the two modalities without the need for additional landmarks. Integrated methods allow marker-free registration, but the technological challenge of incorporating a light microscope inside the vacuum system of an electron microscope restricts its imaging quality.

What makes integrated systems interesting from the registration point of view is that they can provide LM and EM imaging perfectly aligned to each other. In this case, a monomodal registration of one common channel between the integrated LM and any other previously acquired LM imaging will be sufficient to obtain a final match and a generic solution. The goal of miniaturized and adaptive devices such as the miniLM for serial blockface scanning electronic microscopes [9] is to allow both a marker-free registration and a flexible choice of imaging modalities.

Up to now, the mechanical alignment of integrated microscopes has alignment accuracies of about 100 nm [9]. By using cathodoluminescence marker points generated by the electron beam of a SEM microscope, an accurate and automatic registration with an alignment precision of 5 nm can be reached [38, 44]. These landmarks can be generated on the fly and at any position, allowing the creation of a dense coverage in any area of the image. Computer vision-based generation of live landmark patterns could reduce the error to near zero. The idea is to generate luminescence patterns centered on the structure of interest, optimizing the following equation:

$$\left\langle TRE^2 \right\rangle = \left\langle FLE^2 \right\rangle \left(\frac{1}{N} + SUM\left(\frac{di}{fi} \right) \right)$$

where TRE is the target registration error and FLE the fiducial localization error (for example about 100 nm in lm). This means that at the center of gravity of the pattern of the fiducial, the expected average accuracy is the square root of FLE^2/N – even with 100 nm of fiducial localization error, 10 nm of resolution could be reached with 100 points.

11.4.2 Incorporate a Priori Knowledge of the Specimen

In all major problems of bioimage analysis (e.g. segmentation or tracking), it appears that including the maximum amount of information known by the user leads to the best-performing algorithm. For example, in tracking, i.e. following accurately particle or cells, knowing the dynamic type (diffusive or directional for example) greatly improves the estimation of the dynamic parameters [45].

Here, we believe that better modeling of the deformation basis would greatly enhance the accuracy of registration (see also Figure 11.3, which shows that a different non-rigid basis leads to different spatial correlation). Indeed, this is the case when chromatic aberration in LM needs to be corrected for realigning two fluorescence channels: in this case, a distortion model would lead to the best result [46]. Other examples of using the knowledge of the basis of expected deformation have also been shown in the medical image registration literature [43].

Some knowledge has already been generated for the type of deformation to be expected between two modalities of microscopy. In CLEM, for example, most of the deformation occurs at the sample preparation step for EM. While vitrification techniques allow preservation of biological samples in a near-native state with limited physical deformations [1, 34, 47–49], structural changes and sample deformations can occur at several steps when using chemical fixation, freeze substitution, and sectioning [50–52]. This may be due to either the sample preparation: the embedding process (dehydration, embedding, or vitrification [53]); the sectioning or the electron imaging [54, 55]; or the image-acquisition constraints: delay before vitrification [52, 56, 57]); chromatic distortion in LM [34]; or electron beam aberration or damage in EM [58].

Different studies have started to characterize the deformation occurring in the sample during sample preparation. For example, in cryo-sections during freeze drying, the deformation has been shown to be isotropic with the loss of water [59], with a shrinkage of about 10%, and no morphological differences were observed in cellular compartments, meaning that each structure did not change its shape. In chemical fixation, the artefacts and morphological changes are more complex [60].

To our knowledge, no formal study in 3D has been performed.

11.4.3 Toward the Use of Machine Learning

Machine learning, and in particular deep learning, today is the state of the art for image segmentation [61]. We can also expect the same technical improvement for image registration in CLEM. Supervised or semi-supervised machine learning does not create new processing, but automatically selects a set of adapted methods to a particular problem based on a training set, given the expectation of the user. Deep learning has the particularity of creating the tested methods by itself in an abstract way. There are reports about the successful use of deep learning in registration in medical imaging [62, 63], but also in the multimedia field, when fusing data as unpredictable and uncorrelated as sound tracks with images of a video [64]. Machine learning could also help provide an intuitive interface by reducing the fine-tuning or the search for the ideal registration method for one particular workflow of correlative microscopy. Indeed, as in bioimage analysis in general, the user experience is a crucial factor for the adoption of a software.

But even before going to actual machine-learning methods, creating a representative catalog of imaging fiducials will already simplify most of the correlative workflows by removing the step to adapt the image-processing workflow to a specific use of one fiducial marker. For example, in the same way as a sample holder grid may be stored (e.g. in the Shuttle and Find System), artificial bead images could be stored in a database in registration software that would ease their automatic retrieval from both EM and fluorescence image, by pattern recognition.

Going further, machine learning could be used with three different strategies: to transform images or features to appear the same in both modalities, to find the good metric of similarity to use, or even to recognize the deformation itself based on the first superimposition.

The first approach consists of using machine learning to artificially make the two modalities more similar. Machine-learning methods were already used, for example, to try to predict the aspect of an image, using image analogies, to come back to a mono-modal registration approach [33]. However these approaches are more related to the creation of dictionaries or catalogs and may be difficult to apply in the context of sparse fluorescent imaging.

Similar approaches were developed [65] that use deep learning of the features to be used for the registration (or, more precisely, the kind of filter to be applied to emphasize these features), even with huge scale differences as we face them in CLEM. This could be a first step toward a very generic semi-supervised approach where the user manually selects a few points that will be tracked over scales and images based on reproducible descriptors. We can even envision an entirely generic approach where whatever the images and type of features, the algorithm would learn the best features to follow.

Two other possible strategies have already been identified in the medical image registration field [63]: (1) to use deep-learning networks to estimate a similarity measure for two images to drive an iterative optimization strategy; and (2) to directly predict transformation parameters using deep regression networks. None of these approaches have been used yet in CLEM. The second one, in particular, could be of great interest by teaching it not to learn the parameters itself but, rather, the basis of transformation. Indeed, as already noted, one of the difficulties in correlative registration is to take into account descriptions image distortions by specimen deformation or by the imaging process for a very accurate alignment.

All these approaches would require training datasets, i.e., preregistered datasets with a lot of different conditions of imaging and samples. For this, we would need to ask the community to create a database of manually or automatically registered images in correlative microscopy, containing source and target raw data, as well as the transformation computed (which may not be the ground truth). Some public databases are already available, but they tend to focus on one modality, LM [66], [67] or EM [68, 69]. Where and how to host this correlative data is still an open question, but some efforts are now being made, led by the EMBL-EBI and the EMPIAR project [69]. Ideally, the data should be accessible by computer software through an interactive interface, comprising a dictionary of transformation bases and manipulation of parameters to reapply the resulting transformation back and forth. This would allow users to work on the registered image without having to store them and provide training datasets, for machine learning but also other approaches developed by the bioimage informatics community.

Apart from the community effort, crowdsourcing and citizen science through serious games may also help to gather a sufficient amount of data. Similar effort have been started successfully for EM segmentation to train computers to segment automatically [70]. Nonscientific volunteers are asked to segment a nucleus on an EM image. We could imagine a similar game where volunteers could either assess the quality of alignment of images or create transformations by matching points in order to create a dense map of deformation that could be used to learn the good physical modelization.

11.5   Visualization of Correlation

One remaining problem is the data representation and its visualization.

   The natural way of visualizing the correlation is to overlay the two registered images, using different colors/look-up tables or applying transparency. For 2D data, navigating in the images at a very different scale requires saving the different views of the registration. A Google Maps-like user interface, where different modalities could be superimposed on-the-fly at different scales, may help our understanding. Different sources of information could be superimposed and navigated at different views, with the necessary resolution. This can be achieved with a pyramidal representation of data and specific file formats. In addition, in 3D, to ease the correlation, but also the navigation in 3D, virtual reality could help, as, for example, immersive microscopy such as proposed by Arrivis (inViewR). An alternative could also be the 3D holograms that started to be used in clinical practice [71, 72]. Manipulating the volume in 3D, erasing a part of the visualization to focus on one, being able to catch and manipulate the volume manually as if it was a real object would help. However, the density of information in 3D EM data strongly limits its visualization by volume rendering. FM datasets are easier to visualize because the objects of interest are usually on a black background and are much more sparsely distributed. In order to ease the matching in 3D, one of the most direct solutions would require the EM data to be presegmented and visualize surface renderings (at least for the landmarks of interest in the case of automatic registration) of, ideally, all structures of interest. With the progress of machine learning and computer vision in the challenging field of segmentation, this dream may become possible in the near future, at least for well-defined and known structures such as nuclei.

   In addition, while augmented reality is now becoming routine in other fields like aeronautics or medical imaging, i.e. used for surgery for more than 10 years (Figure 11.5), it is not yet used in the field of correlative microscopy. Superimposing the extra information from another imaging modality to the observation at the "target" microscope also requires registration. Similarly to what happens during image-guided-surgery (Figure 11.5), augmented reality could be used to guide the ultramicrotome, using some external fiducials on the resin block or the sample itself, to display the place of the region of interest observed in LM in the ocular of the microtome microscope. Modern devices now even allow 3D and colored projections in the microscope ocular [73, 74].

11.6   Conclusion

CLEM is a powerful but very demanding method, relying on a host of different expertise. Going to high-throughput screening (HTS) by CLEM still requires a lot of progress on different fronts. Some preliminary work has already sped up the different processes for the correlation of images, moving toward automated correlative imaging [16]. We hope that the envisioned progress will also contribute in the future to decreasing the time and effort of using this powerful method of imaging.

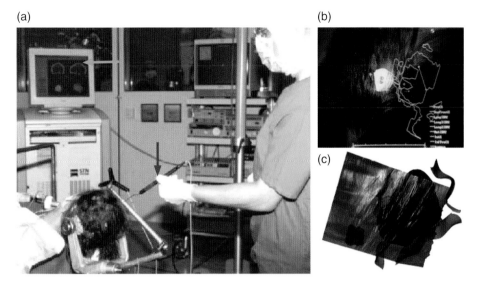

Figure 11.5 Inspiring neurosurgery guiding through augmented reality and augmented virtuality. (a) The neurosurgeon is pointing to the patient, the position of the pointer (red arrow) is simultaneously appearing on the screen (green arrow) showing the preoperative imaging at its position. (b) Capture from the views in the surgical microscope displaying additional information in green: simplified scheme of element of region of interest extracted from the preoperative imaging are displayed live on the patient. (c) A similar view in 3D. Images from [75]. The same methods of visualization could be applied for guiding any manual step in ultramicrotomy, for example, or during the acquisition.

Acknowledgments

The authors are grateful to Bertrand Simon for the recent literature on augmented reality in microscopy, Charles Kervrann and Patrick Bouthemy for fruitful discussion about image registration, Philippe Hulin and Xavier Heiligenstein for discussion on correlation data visualization. Figures were realized using FigureJ [76].

References

1 Schorb M, Briggs JA. Correlated cryo-fluorescence and cryo-electron microscopy with high spatial precision and improved sensitivity. Ultramicroscopy. 2014 143:24–32.

2 Hodgson L, Nam D, Mantell J, Achim A, Verkade P. Retracing in correlative light electron microscopy: where is my object of interest? Methods Cell Biol. 2014 124:1–21.

3 Blazquez-Llorca L, Hummel E, Zimmerman H, Zou C, Burgold S, Rietdorf J, et al. Correlation of two-photon in vivo imaging and FIB/SEM microscopy. J Microsc. 2015 259(2):129–36.

4 Karreman MA, Mercier L, Schieber NL, Solecki G, Allio G, Winkler F, et al. Fast and precise targeting of single tumor cells in vivo by multimodal correlative microscopy. J Cell Sci. 2016 129(2):444–56.

5 Karreman MA, Mercier L, Schieber NL, Shibue T, Schwab Y, Goetz JG. Correlating intravital multi-photon microscopy to 3D electron microscopy of invading tumor cells using anatomical reference points. PLoS One. 2014 9(12):e114448.

6 Arnold J, Mahamid J, Lucic V, de Marco A, Fernandez JJ, Laugks T, et al. Site-Specific Cryo-focused Ion Beam Sample Preparation Guided by 3D Correlative Microscopy. Biophys J. 2016 110(4):860–9.

7 Kolovou A, Schorb M, Tarafder A, Sachse C, Schwab Y, Santarella-Mellwig R. A new method for cryo-sectioning cell monolayers using a correlative workflow. Methods Cell Biol. 2017 140:85–103.

8 Mahamid J, Pfeffer S, Schaffer M, Villa E, Danev R, Cuellar LK, et al. Visualizing the molecular sociology at the HeLa cell nuclear periphery. Science. 2016 351(6276):969–72.

9 Brama E, Peddie CJ, Wilkes G, Gu Y, Collinson LM, Jones ML. ultraLM and miniLM: Locator tools for smart tracking of fluorescent cells in correlative light and electron microscopy. Wellcome Open Res. 2016 1:26.

10 Loussert Fonta C, Leis A, Mathisen C, Bouvier DS, Blanchard W, Volterra A, et al. Analysis of acute brain slices by electron microscopy: a correlative light-electron microscopy workflow based on Tokuyasu cryo-sectioning. J Struct Biol. 2015 189(1):53–61.

11 Huynh T, Daddysman MK, Bao Y, Selewa A, Kuznetsov A, Philipson LH, et al. Correlative imaging across microscopy platforms using the fast and accurate relocation of microscopic experimental regions (FARMER) method. Rev Sci Instrum. 2017 88(5):053702.

12 Cheng D, Shami G, Morsch M, Huynh M, Trimby P, Braet F. Relocation is the key to successful correlative fluorescence and scanning electron microscopy. Methods Cell Biol. 2017 140:215–44.

13 Collman F, Buchanan J, Phend KD, Micheva KD, Weinberg RJ, Smith SJ. Mapping synapses by conjugate light-electron array tomography. J Neurosci. 2015 35(14):5792–807.

14 Edelstein A, Amodaj N, Hoover K, Vale R, Stuurman N. Computer control of microscopes using microManager. Curr Protoc Mol Biol. 2010 Chapter 14:Unit14 20.

15 Mastronarde DN. Automated electron microscope tomography using robust prediction of specimen movements. J Struct Biol. 2005 152(1):36–51.

16 Schorb M, Gaechter L, Avinoam O, Sieckmann F, Clarke M, Bebeacua C, et al. New hardware and workflows for semi-automated correlative cryo-fluorescence and cryo-electron microscopy/tomography. J Struct Biol. 2017 197(2):83–93.

17 Karreman MA, Hyenne V, Schwab Y, Goetz JG. Intravital Correlative Microscopy: Imaging Life at the Nanoscale. Trends Cell Biol. 2016 26(11):848–63.

18 Karreman MA, Ruthensteiner B, Mercier L, Schieber NL, Solecki G, Winkler F, et al. Find your way with X-Ray: Using microCT to correlate in vivo imaging with 3D electron microscopy. Methods Cell Biol. 2017 140:277–301.

19 van der Schaar HM, Melia CE, van Bruggen JA, Strating JR, van Geenen ME, Koster AJ, et al. Illuminating the Sites of Enterovirus Replication in Living Cells by Using a Split-GFP-Tagged Viral Protein. mSphere. 2016 1(4).

20 Paul-Gilloteaux P, Heiligenstein X, Belle M, Domart MC, Larijani B, Collinson L, et al. eC-CLEM: flexible multidimensional registration software for correlative microscopies. Nat Methods. 2017 14(2):102–3.

21 Thevenaz P, Ruttimann UE, Unser M. A pyramid approach to subpixel registration based on intensity. IEEE Trans Image Process. 1998 7(1):27–41.

22 Russell MR, Lerner TR, Burden JJ, Nkwe DO, Pelchen-Matthews A, Domart MC, et al. 3D correlative light and electron microscopy of cultured cells using serial blockface scanning electron microscopy. J Cell Sci. 2017 130(1):278–91.

23 Fukuda Y, Schrod N, Schaffer M, Feng LR, Baumeister W, Lucic V. Coordinate transformation based cryo-correlative methods for electron tomography and focused ion beam milling. Ultramicroscopy. 2014 143:15–23.

24 Kukulski W, Schorb M, Welsch S, Picco A, Kaksonen M, Briggs JA. Correlated fluorescence and 3D electron microscopy with high sensitivity and spatial precision. J Cell Biol. 2011 192(1):111–9.

25 Muller A, Neukam M, Ivanova A, Sonmez A, Munster C, Kretschmar S, et al. A Global Approach for Quantitative Super Resolution and Electron Microscopy on Cryo and Epoxy Sections Using Self-labeling Protein Tags. Sci Rep. 2017 7(1):23.

26 Kukulski W, Schorb M, Welsch S, Picco A, Kaksonen M, Briggs JA. Precise, correlated fluorescence microscopy and electron tomography of lowicryl sections using fluorescent fiducial markers. Methods Cell Biol. 2012 111:235–57.

27 Lopez CS, Bouchet-Marquis C, Arthur CP, Riesterer JL, Heiss G, Thibault G, et al. A fully integrated, three-dimensional fluorescence to electron microscopy correlative workflow. Methods Cell Biol. 2017 140:149–64.

28 Klein S, Staring M, Murphy K, Viergever MA, Pluim JP. elastix: a toolbox for intensity-based medical image registration. IEEE Trans Med Imaging. 2010 29(1):196–205.

29 Nam D, Mantell J, Hodgson L, Bull D, Verkade P, Achim A. Feature-Based Registration For Correlative Light And Electron Microscopy Images. In: IEEE, editor. ICIP 20142014.

30 Heiligenstein X, Paul-Gilloteaux P, Raposo G, Salamero J. eC-CLEM: A multidimension, multimodel software to correlate intermodal images with a focus on light and electron microscopy. Methods Cell Biol. 2017 140:335–52.

31 Preibisch S, Saalfeld S, Schindelin J, Tomancak P. Software for bead-based registration of selective plane illumination microscopy data. Nat Methods. 2010 7(6):418–9.

32 Acosta BMT, Bouthemy P, Kervrann C. A Common Image Representation And A Patch-Based Search For Correlative Light-Electron-Microscopy (Clem) Registration. In: IEEE, editor. ISBI 20162016.

33 Cao T, Zach C, Modla S, Powell D, Czymmek K, Niethammer M. Multi-modal registration for correlative microscopy using image analogies. Med Image Anal. 2014 18(6):914–26.

34 Schellenberger P, Kaufmann R, Siebert CA, Hagen C, Wodrich H, Grunewald K. High-precision correlative fluorescence and electron cryo microscopy using two independent alignment markers. Ultramicroscopy. 2014 143:41–51.

35 Cohen EA, Ober RJ. Analysis of Point Based Image Registration Errors With Applications in Single Molecule Microscopy. IEEE Trans Signal Process. 2013 61(24):6291–306.

36 Fitzpatrick JM, editor Fiducial registration error and target registration error are uncorrelated. SPIE7261 2009 Lake Buena Vista, FL.

37 Fitzpatrick JM, West JB. The distribution of target registration error in rigid-body point-based registration IEEE Transactions on Medical Imaging. 2001 20(9):917–27.

38 Haring MT, Liv N, Zonnevylle AC, Narvaez AC, Voortman LM, Kruit P, et al. Automated sub-5 nm image registration in integrated correlative fluorescence and

electron microscopy using cathodoluminescence pointers. Scientific Reports. 2017 7:43621.

39 DM VANE, Bos E, Pawlak JB, Overkleeft HS, Koster AJ, SI VANK. Correlative light and electron microscopy reveals discrepancy between gold and fluorescence labelling. J Microsc. 2017.

40 Zitová B, Flusser J. Image registration methods: a survey. Image and Vision Computing. 2003 21(11):977–1000.

41 Ferrante E, Paragios N. Slice-to-volume medical image registration: A survey. Med Image Anal. 2017 39:101–23.

42 Viergever MA, Maintz JB, Klein S, Murphy K, Staring M, Pluim JP. A survey of medical image registration - under review. Med Image Anal. 2016 33:140–4.

43 Sotiras A, Davatzikos C, Paragios N. Deformable medical image registration: a survey. IEEE Trans Med Imaging. 2013 32(7):1153–90.

44 Iijima H, Fukuda Y, Arai Y, Terakawa S, Yamamoto N, Nagayama K. Hybrid fluorescence and electron cryo-microscopy for simultaneous electron and photon imaging. J Struct Biol. 2014 185(1):107–15.

45 Chenouard N, Smal I, de Chaumont F, Maska M, Sbalzarini IF, Gong Y, et al. Objective comparison of particle tracking methods. Nat Methods. 2014 11(3):281–9.

46 Kozubek M, Matula P. An efficient algorithm for measurement and correction of chromatic aberrations in fluorescence microscopy. Journal of Microscopy. 2000 200(3):206–17.

47 Johnson E, Seiradake E, Jones EY, Davis I, Grunewald K, Kaufmann R. Correlative in-resin super-resolution and electron microscopy using standard fluorescent proteins. Sci Rep. 2015 5:9583.

48 Heiligenstein X, Heiligenstein J, Delevoye C, Hurbain I, Bardin S, Paul-Gilloteaux P, et al. The CryoCapsule: simplifying correlative light to electron microscopy. Traffic. 2014 15(6):700–16.

49 Nixon SJ, Webb RI, Floetenmeyer M, Schieber N, Lo HP, Parton RG. A single method for cryofixation and correlative light, electron microscopy and tomography of zebrafish embryos. Traffic. 2009 10(2):131–6.

50 Korogod N, Petersen CC, Knott GW. Ultrastructural analysis of adult mouse neocortex comparing aldehyde perfusion with cryo fixation. Elife. 2015 4.

51 Mobius W, Nave KA, Werner HB. Electron microscopy of myelin: Structure preservation by high-pressure freezing. Brain Res. 2016 1641(Pt A):92–100.

52 Murk JL, Posthuma G, Koster AJ, Geuze HJ, Verkleij AJ, Kleijmeer MJ, et al. Influence of aldehyde fixation on the morphology of endosomes and lysosomes: quantitative analysis and electron tomography. J Microsc. 2003 212(Pt 1):81–90.

53 Dubochet J, Adrian M, Chang JJ, Homo JC, Lepault J, McDowall AW, et al. Cryo-electron microscopy of vitrified specimens. Q Rev Biophys. 1988 21(2):129–228.

54 Richter K. Cutting artefacts on ultrathin cryosections of biological bulk specimens. Micron. 1994 25(4):297–308.

55 McDonald K. Cryopreparation methods for electron microscopy of selected model systems. Methods Cell Biol. 2007 79:23–56.

56 Verkade P. Moving EM: the Rapid Transfer System as a new tool for correlative light and electron microscopy and high throughput for high-pressure freezing. J Microsc. 2008 230(Pt 2):317–28.

57 Koning RI, Faas FG, Boonekamp M, de Visser B, Janse J, Wiegant JC, et al. MAVIS: an integrated system for live microscopy and vitrification. Ultramicroscopy. 2014 143:67–76.

58 Luther PK, Lawrence MC, Crowther RA. A method for monitoring the collapse of plastic sections as a function of electron dose. Ultramicroscopy. 1988 24(1):7–18.

59 Casanova G, Nolin F, Wortham L, Ploton D, Banchet V, Michel J. Shrinkage of freeze-dried cryosections of cells: Investigations by EFTEM and cryo-CLEM. Micron. 2016 88:77–83.

60 Kellenberger E, Johansen R, Maeder M, Bohrmann B, Stauffer E, Villiger W. Artefacts and morphological changes during chemical fixation. J Microsc. 1992 168(Pt 2): 181–201.

61 Dai W, Fu C, Raytcheva D, Flanagan J, Khant HA, Liu X, et al. Visualizing virus assembly intermediates inside marine cyanobacteria. Nature. 2013 502(7473): 707–10.

62 Shen D, Wu G, Suk HI. Deep Learning in Medical Image Analysis. Annu Rev Biomed Eng. 2017 19:221–48.

63 Geert Litjens TK, Babak Ehteshami Bejnordi, Arnaud Arindra Adiyoso Setio, Francesco Ciompi, Mohsen Ghafoorian, Jeroen A.W.M. van der Laak, Bram van Ginneken, Clara I. Sánchez. A Survey on Deep Learning in Medical Image Analysis. 2017.

64 Vukotić V, Raymond C, Gravier G. Multimodal and Crossmodal Representation Learning from Textual and Visual Features with Bidirectional Deep Neural Networks for Video Hyperlinking ACM Multimedia 2016 Workshop: Vision and Language Integration Meets Multimedia Fusion (iV&L-MM'16) Amsterdam, Netherlands 2016.

65 Wu G, Kim M, Wang Q, Munsell BC, Shen D. Scalable High-Performance Image Registration Framework by Unsupervised Deep Feature Representations Learning. IEEE Trans Biomed Eng. 2016 63(7):1505–16.

66 Eliceiri KW, Berthold MR, Goldberg IG, Ibanez L, Manjunath BS, Martone ME, et al. Biological imaging software tools. Nat Methods. 2012 9(7):697–710.

67 Williams E, Moore J, Li SW, Rustici G, Tarkowska A, Chessel A, et al. Image Data Resource: a bioimage data integration and publication platform. Nat Meth. 2017 advance online publication.

68 Patwardhan A, Carazo JM, Carragher B, Henderson R, Heymann JB, Hill E, et al. Data management challenges in three-dimensional EM. Nat Struct Mol Biol. 2012 19(12):1203–7.

69 Iudin A, Korir PK, Salavert-Torres J, Kleywegt GJ, Patwardhan A. EMPIAR: a public archive for raw electron microscopy image data. Nat Methods. 2016 13(5):387–8.

70 Collinson L, Martin J, Peddie C, Strange A, Weston A, Hutchings R, et al. Etch a cell https://www.zooniverse.org/projects/h-spiers/etch-a-cell2017

71 Bruckheimer E, Rotschild C, Dagan T, Amir G, Kaufman A, Gelman S, et al. Computer-generated real-time digital holography: first time use in clinical medical imaging. Eur Heart J Cardiovasc Imaging. 2016 17(8):845–9.

72 Addetia K, Lang RM. The future has arrived. Are we ready? Eur Heart J Cardiovasc Imaging. 2016 17(8):850–1.

73 Huang Y-H, Yu T-C, Tsai P-H, Wang Y-X, Yang W-L, Ouhyoung M. Scope+: a stereoscopic video see-through augmented reality microscope. SIGGRAPH Asia 2015 Emerging Technologies Kobe, Japan. 2818476: ACM 2015. p. 1–3.

74 Zhang X, Chen G, Liao H. High Quality See-Through Surgical Guidance System Using Enhanced 3D Autostereoscopic Augmented Reality. IEEE Trans Biomed Eng. 2016.

75 Paul-Gilloteaux P. Neurochirurgie guidée par l'image: visualisation mixte et quantication des déformations cérébrales peropératoires à l'aide de reconstructions stéréoscopiques de la surface corticale: Université de Rennes 1 2006.

76 Mutterer J, Zinck E. Quick-and-clean article figures with FigureJ. J Microsc. 2013 252(1):89–91.

12

Big Data in Correlative Imaging

Ardan Patwardhan[1] and Jason R. Swedlow[2]

[1] European Molecular Biology Laboratory, European Bioinformatics Institute (EMBL-EBI), Wellcome Genome Campus, Hinxton, United Kingdom
[2] Centre for Gene Regulation and Expression, University of Dundee, United Kingdom

12.1 Introduction

Correlative light-electron microscopy (CLEM) today spans a wide range of scales, from the molecular domain of structural biology all the way to organism-level imaging and involves a diversity of imaging modalities. The field is growing rapidly, with technological and advances in sample preparation, imaging and signal detection, and data processing and analysis. Previous experience in molecular and structural biology shows that important synergies and coherence in rapidly emerging fields can be achieved by building public data archives, where the datasets associated with scientific publications can be stored, shared, accessed, and compared.

To support the multimodal nature of correlative imaging experiments we need public archiving for all the relevant imaging modalities as well as the means to link all this information together. All of the public data resources that now serve as the foundation for the modern biosciences evolved through several steps, and many started as quite limited resources and grew in scale and functionality as technology, expertise, and capabilities steadily increased. We expect the same type of progression to occur across the various imaging data resources that are now appearing, and in particular for CLEM data resources.

In order to gain a perspective on the construction of public data resources for CLEM data, we start by recalling how public archiving in structural biology has evolved from mainly catering for molecular structure to efforts to broaden the scope to include cellular structure. We then take a look at how systems imaging biology has become a compelling use-case for the benefits of the public archiving of imaging data and has led to the first efforts to set up public archiving for bioimaging data.

12.2 The Protein Data Bank

The Protein Data Bank (PDB), set up in 1971 as a public archive for experimentally derived atomic coordinate models for biological macromolecules and complexes, is not only one of the oldest surviving digital biological public archives but also a thriving and flourishing one that has grown to encompass over 140000 entries from X-ray crystallography, nuclear magnetic resonance (NMR), and electron microscopy (EM) experiments (H. M. Berman et al. 2012). Over half a billion international users access PDB data annually for a variety of purposes, including using the data to bootstrap the determination of new structures, comparative analysis with other structures, the development of generalized structural concepts (e.g., structural classification schemes such as CATH (http://www.cathdb.info/) (Dawson et al. 2017) and SCOP2 (http://scop2.mrc-lmb.cam.ac.uk/) (Andreeva et al. 2014), data-mining, validation, methods development, teaching, and training. A variety of web-based visualization and analytical tools have been developed that make this data accessible to a range of users from the wider scientific community. The PDB is managed by the members of the Worldwide Protein Databank (wwPDB) (H. Berman et al. 2007) – the Research Collaboratory for Structural Bioinformatics PDB (RCSB PDB), the Protein Data Bank in Europe (PDBe) at the European Bioinformatics Institute (EMBL-EBI), the Protein Data Bank Japan (PDBj) at the Institute for Protein Research in Osaka University, and the Biological Magnetic Resonance Bank (BMRB) at the University of Wisconsin-Madison. Beyond the immediate benefits of data reuse, the people and organizations involved in running the PDB play a key role as community facilitators on vital issues such as deposition policy, data models and formats, and raising awareness and improving the state of validation in the field. To get to this stage of buy-in and support from the structural biology community as to the merits of public archiving has taken many years and patience and perseverance on the part of the people involved, and has also required engagement with the funding agencies and journals. It took until the 1990s for the deposition of coordinates to become mandatory upon publication to most relevant journals (Kleywegt et al. 2004).

12.3 Resources for Cryo-EM

Inspired by the role PDB had played for the structural biology community, there were growing calls in the early 2000s from the cryo-electron microscopy (cryo-EM) community to similarly set-up public archiving for the 3D reconstructions that were derived from cryo-EM experiments. This led to the establishment of the Electron Microscopy Data Bank (EMDB) at EMBL-EBI in 2002 (Tagari et al. 2002). EMDB now holds over 6000 entries from a variety of EM techniques, including single-particle, electron tomography, sub-tomogram averaging, and 2D and 3D electron diffraction experiments (Patwardhan 2017). The pace of depositions to EMDB has been accelerating owing to technological advances in the field, such as the introduction of the direct electron detector, which have greatly facilitated the determination of molecular structures to resolutions that were previously the preserve of the more established structural biology techniques such as X-ray crystallography and NMR. It should be noted that while the current holdings of the PDB far exceed that of the EMDB, when it comes to total data volume the reverse is true – the PDB is about 1 GB while EMDB is about 1 TB. These numbers should be contrasted with the requirements for bioimaging archiving where a single dataset can easily lie in the tens of terabytes range.

In 2011, we organized the "Data Management Challenges in 3D Electron Microscopy" workshop where one of the key outcomes was the need for public archiving of the 2D raw image data related to the 3DEM reconstructions in EMDB for purposes of validation, methods development, training etc (Patwardhan et al. 2012). A key point to note here is that similar to bioimaging data, a raw cryo-EM image dataset can easily lie in the tens of terabytes range. The infrastructure surrounding EMDB was not built to cope with the storage, transfer and processing of such large amounts of data. Therefore in 2014 following the successful acquisition of funding from the MRC and BBSRC, the Electron Microscopy Public Image Archive (EMPIAR (Iudin et al. 2016); described in more detail below) was set up. The scope (molecular to cellular) of electron tomography experiments that could be deposited to EMDB was also discussed and there was a general consensus that no arbitrary restrictions should be placed as it would be difficult to define clear rules for doing so and because of the growing value of cellular electron tomography. Whether the deposition of electron tomography reconstructions should be made mandatory was also discussed. While there was a consensus with regards to doing so for sub-tomogram averaging experiments (which would typical be of molecular assemblies), the best that could be achieved for the electron tomograms themselves was a strong "recommendation" to deposit a representative tomogram. One of the key arguments put forward by those opposing the deposition requirements was that given the amount of work they needed to do to get the tomograms in the first place and the richness of information in them, which they had quite often not explored fully by the time of their first publications, they wanted to have the opportunity to explore this data themselves first, and this could require years.

There is no one specific date when the relevant journals started making deposition mandatory, but an important event in this process was a discussion at the 2012 3DEM Gordon Research Conference to define a deposition policy, followed by a letter-writing campaign to journals. It is interesting to note that it took "only" 10 years for EMDB to reach this stage compared with decades for PDB. Undoubtedly, the fact that both archives involve overlapping domains of expertise allowed some of the hindsight and experience gained with PDB to be easily adopted to EMDB.

With a growing number of structural biology experiments involving a combination of techniques and data from different scales, we organized an expert workshop on *"A 3D Cellular Context for the Macromolecular World"* in 2012 (Patwardhan et al. 2014). The workshop participants agreed that it would be unrealistic to consider all scales of imaging data at once and that instead it would be useful and valuable for the scientific community to begin archiving the modalities at the molecular to cellular scales, viz. 3D scanning electron microscopy (3DSEM), soft X-ray tomography (SXT) and the light microscopy (LM) data from correlative light and electron microscopy (CLEM) experiments. Soon after EMPIAR was set up in 2014, its remit was expanded to allow for the deposition of 3DSEM and SXT data albeit with a very primitive data model. It was still unclear at this stage as to where the LM data from CLEM experiments would go and how this could be linked to the EM data.

Another important topic discussed at this meeting was the need to be able to integrate (link) cellular imaging data with other structural and imaging data, and with other bioinformatics resources. Without such integration, the wider reuse of cellular imaging data would be severely restricted. These considerations led to the expert workshop on *"3D segmentations and transformations – building bridges between cellular and molecular structural biology"* in 2015, bringing together a range of expertise including microscopists, developers of segmentation software and ontology experts (Patwardhan et al. 2017). To integrate cellular structure data, we need segmentation data (i.e. 3D regions of interest, ROIs) that is annotated with biologically relevant terms from established ontologies and bioinformatics resources. Transformations are another essential means for data integration between different scales and imaging modalities – to spatially relate sub-tomograms or atomic coordinate models to their parent EM maps or tomograms, and also to relate images of the same sample acquired using different modalities (e.g. EM and LM). There are no widely supported formats for describing segmentations with structured biological annotations or transformations, and this is a major impediment to data integration. An outcome of this meeting was the development of a file format to represent segmentations and transformations with structured biological annotation – the EMDB Segmentation File Format (EMDB-SFF; https://www.ebi.ac.uk/pdbe/emdb/emschema.html/) and the development of a Python toolkit to enable inter-conversion between various segmentation file formats and EMDB-SFF. In order to facilitate the process of biologically annotating segmentations, we are developing a web-based segmentation annotation tool and a web-based volume browser (back-ended by OMERO) that can provide an integrated view on structural and imaging data from linked datasets.

As mentioned previously, while 3DSEM and SXT could be deposited to EMPIAR, the data model used was relatively primitive and did not capture relevant experimental information. We organized an expert workshop in 2017 *"Expert workshop on public archiving of cellular Electron microscopy and soft X-ray tomography data"* to help establish the archiving requirements for these imaging modalities and data models for EMPIAR. This meeting provided concrete feedback on the experimental data that needed to be captured. The meeting also, once again, highlighted the need to capture the correlative LM data that was so often a part of these experiments.

12.4 Light Microscopy Data Resources

By contrast to the situation in EM, the development of public resources for LM has followed a much less direct path. LM encompasses a very broad range of applications and research domains and in the last 10 years, has undergone an absolute revolution in the number of imaging modalities, research applications and range of resolution available. For these reasons it has been much harder to conceive, let alone build, a single data resource for LM data. Notwithstanding this diversity, it is quite clear that, at least in the foreseeable future, the number of domains of LM that are relevant CLEM applications is relatively limited. While this bounding of the data problem for CLEM simplifies requirements, it is likely that CLEM data resources will reuse technologies and expertise developed for LM data instead of building de novo solutions.

The diversity of LM methods and data has meant that the development of the tools the technology and metadata standards for LM has not been driven by a specialized resource like the EMDataBank but instead has been taken forward by several projects that are building software tools for handling light microscopy imaging data. One of the leading projects is the open microscopy environment (OME: https://openmicroscopy. org), which has been building metadata specifications, open file formats, and open-source software for light microscopy since 2002 (Swedlow et al. 2003). OME's metadata specification, the OME Data Model (Goldberg et al. 2005) has been incorporated into an open file format, OME-TIFF, which is now used in many different academic and commercial software packages. OME has also released open source software platforms that use these specifications as interfaces to complex, multi-dimensional imaging data. Bio-Formats is a software plug-in that reads proprietary image formats and converts the binary data and metadata into the OME Data Model (Linkert et al. 2010). Any software that uses Bio-Formats can read >150 different formats that it supports. OMERO is an enterprise data management platform that allows users to access, process, share, and publish multidimensional image data (Allan et al. 2012; Burel et al. 2015; Li et al. 2016). Together, Bio-Formats and OMERO allow the integration of imaging data from many different applications, annotation with experimental, analytic and phenotypic metadata associated with imaging experiments, enterprise-level data processing, and the controlled or public sharing of data as appropriate.

Several groups have used the capabilities of Bio-Formats and OMERO to build image data publication systems. These include the ASCB CELL Image Library (Orloff et al. 2013), the JCB DataViewer (https://jcb-dataviewer.rupress.org), the SSBD (Tohsato et al. 2016), as well as several institutional repositories associated with imaging projects. Most recently, Bio-Formats and OMERO have been used to build the Image Data Resource, an online database of reference images associated with scientific publications (Williams et al. 2017). IDR integrates datasets from several different types of imaging application across independent studies and links these through annotations of genes, small molecules, and phenotypic data. Where appropriate, IDR links imaging data and metadata to defined external resources like external molecular and model organism databases. Thus, IDR can be a candidate for LM data related to correlative images where the electron microscopy component of the dataset is stored in, for example, EMDB and EMPIAR.

Next we take a closer look at EMPIAR and IDR. We then review the current state of public archiving for bioimaging data, plans that are in the pipeline, and discuss the challenges we see in broadening the scope of public archiving for bioimaging data and in promoting the reusability of correlative imaging data.

12.5   EMPIAR

EMPIAR was originally set up to accept raw image data related to EMDB entries. This rule was put in place to ensure that EMPIAR did not end up becoming a cryoEM image store for the whole world – a situation that would be difficult to justify in terms of return on investment to our paymasters and difficult to cope with from a technical and practical perspective, especially when the archive was just being set up. On the other hand, to encourage deposition, given that there were no relevant journals mandating deposition, we took a non-prescriptive approach to the definition of "raw image data," allowing multi-frame micrographs, frame-averaged micrographs, particle stacks (and combinations of these categories), and tilt series to be deposited. We also allowed deposition of data in different image formats. The first depositions thus became a learning process allowing us to discover what data the community found appropriate to deposit and what type of data EMPIAR users found most useful to download.

At the outset, our expectation was that depositors would prefer to upload as little data as possible and would therefore opt for depositing the cleaned-up particle-stacks only, and similarly that users would preferably download these stacks as they were the smallest. However, owing to the fact that EMPIAR was being set up at approximately the same time as the use of direct electron detectors was starting to take off, and algorithms for the alignment and averaging of multiframe micrographs was a hot topic for development, many users preferred to download the data in its most raw form, which in some cases involved downloading datasets that were terabytes in size. Many depositors were willing to upload not only the multiframe micrographs but also comprehensive sets, including particle stacks and auxilliary files, making EMPIAR a key distribution point for their data and also lessening the burden for hot-storage of this data at their local institutions.

In the past few years, the remit of EMPIAR has been expanded to also allow the deposition of 3DSEM and SXT data. As the 3D structures from these imaging methods cannot be deposited to EMDB, the requirement for these modalities is that they must be directly related to a journal publication, or be a part of a community dataset (e.g. for benchmarking).

EMPIAR today consists of 150+ entries totaling more than 60 TB. Over 14 of these datasets are larger than 1 TB in size, and one is over 12 TB. EMPIAR data is actively used by the community for aiding new scientific work, methods development, validation, and training, and a search of Europe PMC for "EMPIAR" yields over 70 related publications.

EMPIAR now has a comprehensive ecosystem of functionality accessible from a web browser (empiar.org) including a web-based deposition system, data search, and download capability. Every EMPIAR entry has an individual page that provide more information on the entry, offering the possibility to view down-sampled versions of the image data (images may often be over a GB in size) and select parts of a dataset to download.

Going forward, two key areas are the development of data pipelines from microscopy centres such as eBIC (Clare et al. 2017) to EMPIAR in order to streamline and facilitate the deposition of data, and the improvement of the data model to better capture cellular imaging experiments. An open but vital question with regards to cellular imaging data in EMPIAR is where the LM data from correlative experiments will be archived and how the relationships between the different imaging modalities (e.g. transform matrices) to bring them into register will be represented and captured.

12.6 IDR: A Prototype Image Data Resource

To demonstrate the capability and utility of publishing LM image data, we have built the Image Data Repository (IDR), populated with community-submitted image datasets, experimental and analytical metadata, and phenotypic annotations (Williams et al. 2017). The first version of this resource is built and deployed in on EMBL-EBI's Embassy cloud resource (https://idr.openmicroscopy.org). IDR currently holds ~50 TB of image data in ~38 Mio images, and includes all associated experimental (e.g. genes, RNAi, chemistry, geographic location), analytic (e.g. submitter-calculated regions and features), and functional annotations. Wherever possible, metadata in IDR links to external resources that are the authoritative resource for that metadata (Ensembl, NCBI, PubChem, etc.). Datasets in human cells (e.g., http://goo.gl/1zoIIk), Drosophila (http://goo.gl/jPfM3j), and fungi (e.g., http://goo.gl/yFPQCw; http://goo.gl/n3ix5v) are included. The full Mitocheck dataset (http://goo.gl/2FfBwd) and the Tara Oceans, a global survey of plankton and other marine organisms are included (http://goo.gl/2UWWnj).

Wherever possible, functional annotations (e.g. increased peripheral actin), have been converted to defined terms in the EFO, CMPO or other ontologies, always in collaboration with the data submitters (e.g. http://goo.gl/mvKarG). More than 80% of the functional annotations have links to defined, published controlled vocabularies. IDR provides a user-friendly unified interface that supports searches for genes (https://goo.gl/wivV3i), small molecules (https://goo.gl/ntQsbA), and phenotypes (https://goo.gl/Va8vnr). IDR contains imaging data from super-resolution, high content screening, time lapse, light sheet fluorescence microscopy, histological whole-slide imaging, and mouse and human clinical datasets.

The integration of image-based phenotypes and calculated features makes IDR an attractive candidate for computational re-analysis. To make IDR's TB-scale datasets truly available for reuse, we have connected IDR to a JupyterHub-based computational resource (https://idr.openmicroscopy.org/jupyter) that exposes IDR datasets via a public API (https://idr.openmicroscopy.org/about/api.html). Exemplar notebooks are available that provide analysis of image features using PCA, querying of images by annotated by phenotypes, and several others (https://github.com/IDR/idr-notebooks). Users can also use their own notebooks to run their own analyses.

Given the diversity of LM technologies and applications, it is unlikely that a single data public resource can integrate all LM data or derive value from doing so. Therefore, it is likely that multiple imaging resources, each targeting different domains, applications, or purposes will emerge. To encourage reuse of IDR technology in LM data resources, we have made the software systems that power IDR as generic as possible. For instance, Bio-Formats reads imaging data from a range of imaging modalities and OMERO's metadata structures and APIs are flexible and are used across many different domains. Moreover, the integrated IDR application stack, including all IDR databases, search tools and APIs are available for download and can be deployed via Ansible scripts that automate the deployment of the IDR software stack (https://github.com/IDR/deployment).

12.7 Public Resources for Correlative Imaging

12.7.1 CLEM Data Formats

One of the inevitable challenges for any public data archiving is the file formats used for data submission. As noted above, a CLEM format does not have to support all of the domains of light microscopy, so it is in principle easier to devise a candidate community format for CLEM data than for all of LM or EM. However, several commercial manufacturers have entered the CLEM field, and each of these have developed proprietary custom formats for storing the data produced by their CLEM systems. Therefore, it is likely that CLEM datasets will be submitted to a public resource, at least in the short term, in many different formats.

The specification of a CLEM data format requires consideration of the aspects of CLEM imaging that are similar to other modalities and those that are specific or unusual. Common imaging metadata includes pixel size, image dimensions, pixel bit depth, and endianness. However, CLEM formats must include metadata that describe the spatial relationships between the different imaging modalities used in the experiment. These include transformation matrices, warping parameters, and any defined ROIs and segmented areas that relate the different imaging modalities to one another. Moreover, these metadata have to account for the different dimensionality of the LM and EM components of a CLEM experiment. For example, a CLEM experiment might include a time-lapse LM sequence where, at specific timepoint, the sample is flash-frozen and prepared for EM and then imaged. Alternatively, the LM experiment may only involve a single 2D image that correlates to a 2D position in a 3D tomogram or 3D scanning EM data stack. In each of these cases, the metadata that relates the LM and EM images must also be recorded to ensure faithful alignment.

All of these parameters must be stored in a defined and open CLEM format, which, in the longer term, a public CLEM data repository might help define and implement. In fields like molecular and structural biology, public repositories have helped catalyze the development of open standardized file formats for data submission by members of the community (e.g. the PDB format). As these formats became used for submission, their adoption grew and they evolved into community-driven standards. The central role of the public repository in developing, supporting, and ultimately requiring a community standard for data submission is another argument for the construction of a public CLEM data resource.

12.8 Future Directions

12.8.1 A BioImage Archive

With the growing success of EMPIAR and IDR in dealing with cellular imaging data there were growing calls from organizations such as Euro-BioImaging to take a more comprehensive view on the issue of public archiving of bioimaging data. An expert workshop on this topic was held at EMBL-EBI in January 2017 (Ellenberg et al. 2018), which proposed the creation of a large, centralized imaging back-end repository, the Bio-Image Archive (BIA), with storage that could scale into the Exabyte range and could be used as the back-end not only by the existing imaging archives such as EMPIAR and IDR but also for archives that, in the future, would cater to a wider range of bioimaging modalities.

The concept of having a large single back-end store with multiple data ingest pathways (e.g. EMPIAR and IDR) and multiple data-out resources (e.g. EMPIAR and IDR) presenting the data to users was partially inspired by the storage architecture used for sequence data, where the European Nucleotide Archive (ENA) serves as the large back-end store. The architecture consists of wrapping a highly efficient and scalable object store core with an interface layer developed at EMBL-EBI to facilitate access to the data. A pilot study is currently underway to see if this storage infrastructure can be repurposed for BIA. At the same time, EMBL-EBI is actively seeking dedicated funding for storage for BIA.

12.8.2   CLEM Data Submission Pipelines

While the BIA provides one potential solution for the back-end storage problem for the public archiving of imaging data, several challenges must be overcome before there is a viable solution for the public archiving of CLEM and other correlative imaging data. One of the biggest challenges is establishment of usable submission systems that are simple enough that data submission can be a routine part of scientific publication.

A balance needs to be struck between capturing enough information to make the data reusable while not asking so much as to annoy prospective users to the point where they are dissuaded from filling out the information seriously. In general, manual form filling is an error-prone process, and a dialogue is needed with instrument manufacturers to encourage them to write out information relating to the experiment in files that can be harvested during deposition.

As already noted, CLEM has specific metadata that are essential for making proper use of the data. For example, proper alignment and use of CLEM data requires the transformation matrices that relate LM and EM imaging datasets. For many experiments, affine matrices would be sufficient, but especially for methods that involve mechanical sectioning, ways to describing warping must also be considered. Currently, there are no commonly accepted standards for describing transformation matrices, and several different formats are used in LM (e.g. BigDataViewer) and EM (e.g. EMDB-SFF). Neither of these are community-accepted standards yet.

Finally, spatial alignment of data from different imaging modalities may require several images at intermediate resolutions. For example, the LM image may correlate with a low magnification EM image, which is then successively aligned to EM images at high magnifications. Formats for calculation and annotation of intermediate resolution are not yet standardized and are another opportunity where a defined submission format might become a community standard.

12.8.3   Scaling Data Volumes and Usage

The prospect of data expanding faster that the available storage is an ever-present bogeyman and difficult to address with absolute certainty. However, in our experience, the effort that will be required to cement community buy-in in the bioimaging community and the natural inertia in getting scientists to make their data public mean that the organic growth of correlative imaging data in public archives will be a slow process. However, given the capacity of multidimensional LM to generate very large volumes of data in short periods of time, it is quite possible that the limits of available storage will be tested. Major efficiencies could be achieved, as has been the case for nucleic acid databases, by employing efficient compression techniques.

As the public archives for imaging data grow, there will be an increasing need to perform analytics across large numbers of datasets to extract new information that could not be obtained from individual experiments alone. One example would be to reanalyze the distribution or phenotypic characteristics of a particular subcellular structure or region across different types of cells. The viability of such analysis could be severely hampered if it requires the download of large image datasets, especially if the data is at different resolutions and in different formats and frames of reference. It will therefore be essential to develop views to the image data built on top of the back-end store that present data to users in a coherent and unified manner, providing multiscalar representations and efficient access to subvolumes so that users need only download pertinent data at the required resolution. If the problem of representing the transformations required to bring the different modalities into register has been solved, then the presence of correlative imaging data could substantially facilitate such cross-dataset analysis, as the correlative information could be used to pinpoint the pertinent regions for analysis. This capability will allow the construction of added-value databases or knowledge bases that can consume publicly archived data and deliver advanced processing, analysis and integration functions in a cloud-based environment. This approach has been built in IDR for LM data and could be built for CLEM data. This would provide another opportunity to demonstrate the value of publishing and reusing CLEM data.

Furthermore, if computational resources can be brought closer to the data in a cloud, for instance, in a cloud environment (as has been tested with IDR), there is no longer even a need to download the data.

12.8.4 Community Adoption and International Engagement

Unlike the structural community where the concept of public archiving has grown gradually since the 1970s to the point where deposition is now almost considered a natural part of the process, the understanding of the need for and the benefits of public archiving are relatively weak in the bioimaging field. Therefore, a long-term perspective and a great deal of work will be required from the organizations and people involved, and will involve close engagement with the community and other stakeholders such as journals and funding agencies. Targeting high-value CLEM datasets that are useful for a wide range of users would be a good way to prime the effort and thus raise awareness of the resources.

Funding agencies and governments around the globe are developing policies for the preservation and accessibility of data derived from research funded by them. These initiatives may lead to other initiatives to archive and make image data available to the public. Coordination and harmonization among these initiatives and with the BIA will be essential to ensure that potential depositors are not presented with a fragmented and confusing landscape of archiving resources. Similarly, from a data-out perspective, it will be important that links between related datasets are maintained across the archives and there is a way for users to search across the archives.

Acknowledgments

Work on EMDB and EMPIAR is and has been supported by the US National Institutes of Health National Institute of General Medical Sciences (grant R01 GM079429), the UK Medical Research Council with co-funding from the UK Biotechnology and

Biological Sciences Research Council (MRC/BBSRC; grants MR/L007835 and MR/P019544), the BBSRC (grants BB/M018423 and BB/P026893), the Wellcome Trust (grants 088944 and 104948), the European Commission Framework 7 Programme (grant 284209), and EMBL-EBI.

References

Allan, C., Jean-Marie, B., Moore, J., Blackburn, C., Linkert, M., Loynton, S., et al. 2012. OMERO: Flexible, model-driven data management for experimental biology. *Nature Methods* 9(3):245–53.

Andreeva, A., Howorth, D., Chothia, C., Kulesha, E., and Murzin, A.G. 2014. SCOP2 Prototype: A new approach to protein structure mining. *Nucleic Acids Research* 42 (Database issue): D310–14.

Berman, H., Henrick, K., Nakamura, H., and Markley, J. 2007. The Worldwide Protein Data Bank (wwPDB): Ensuring a Single, Uniform Archive of PDB Data. *Nucleic Acids Research* 35(Database issue):D301–3.

Berman, H.M., Kleywegt, G.J., Nakamura, H., and Markley, J.L. 2012. The Protein Data Bank at 40: Reflecting on the Past to Prepare for the Future. *Structure* 20(3):391–96.

Burel, J-M., Besson, S., Blackburn, C., Carroll, M., Ferguson, R.K., Flynn, H., et al. 2015. Publishing and sharing multi-dimensional image data with OMERO. *Mammalian Genome: Official Journal of the International Mammalian Genome Society* 26 (9–10): 441–47.

Clare, K., Siebert, C.A., Hecksel, C., Hagen, C., Mordhorst, V., Grange, M., et al. 2017. Electron bio-imaging centre (eBIC): The UK National Research Facility for Biological Electron Microscopy. *Acta Crystallographica. Section D, Structural Biology* 73 (Pt 6): 488–95.

Dawson, N.L., Lewis, T.E., Das, S., Lees, J.G., Lee, D., Ashford, P., et al. 2017. CATH: An expanded resource to predict protein function through structure and sequence. *Nucleic Acids Research* 45 (D1): D289–95.

Ellenberg, J., Jason, R., Swedlow, M.B., Cook, C.E., Patwardhan, A., Brazma, A., et al. Public archives for biological image data. 2018 *arXiv [q-bio.QM]*. arXiv. http://arxiv.org/abs/1801.10189.

Goldberg, I.G., Allan, C., Burel, J-M., Creager, D., Falconi, A., Hochheiser, H., et al. 2005. The Open Microscopy Environment (OME) Data Model and XML File: Open Tools for Informatics and Quantitative Analysis in Biological Imaging. *Genome Biology* 6 (5): R47.

Iudin, A., Korir, P.K., Salavert-Torres, J., Kleywegt, G.J., and Patwardhan, A. 2016. EMPIAR: A public archive for raw electron microscopy image data. *Nature Methods* 13 (5): 387–88.

Kleywegt, G.J., Harris, M.R., Zou, J.Y., Taylor, T.C., Wählby, A., and Jones, T.A. 2004. The Uppsala Electron-Density Server. *Acta Crystallographica. Section D, Biological Crystallography* 60 (Pt 12 Pt 1): 2240–49.

Linkert, M., Rueden, C.T., Allan, C., Burel, J-M., Moore, W., Patterson, A., et al. 2010. Metadata Matters: Access to Image Data in the Real World. *The Journal of Cell Biology* 189 (5). Rockefeller University Press: 777–82.

Li, S., Besson, S., Blackburn, C., Carroll, M., Ferguson, R.K., Flynn, H., Gillen, K., et al. 2016. Metadata Management for High Content Screening in OMERO. *Methods* 96 (March): 27–32.

Orloff, D.N., Iwasa, J.H., Martone, M.E., Ellisman, M.H., and Kane, C.M. 2013. The cell: An Image Library-CCDB: A curated repository of microscopy data. *Nucleic Acids Research* 41 (D1). Oxford University Press: D1241–50.

Patwardhan, A. Trends in the Electron Microscopy Data Bank (EMDB). *Acta Crystallographica. Section D, Structural Biology* 73 (Pt 6): 503–8.

Patwardhan, A., Ashton, A., Brandt, R., Butcher, S., Carzaniga, R., Chiu, W., et al. 2014. A 3D Cellular Context for the Macromolecular World. *Nature Structural & Molecular Biology* 21 (10): 841–45.

Patwardhan, A., Brandt, R., Butcher, SJ., Collinson, L., Gault, D., Grünewald, K., et al. 2017. Building bridges between cellular and molecular structural biology. *eLife* 10.7554/eLife.25835.

Patwardhan, A., Carazo, J-M., Carragher, B., Henderson, R., Heymann, J.B., Hill, E., et al. 2012. Data management challenges in three-dimensional EM. *Nature Structural & Molecular Biology* 19 (12): 1203–7.

Swedlow, J.R., Goldberg, I., Brauner, E., and Sorger, P.K. 2003. Informatics and quantitative analysis in biological imaging. *Science* 300 (5616): 100–102.

Tagari, M., Newman, R., Chagoyen, M., Carazo, JM., and Henrick, K. 2002. New electron microscopy database and deposition system. *Trends in Biochemical Sciences* 27 (11): 589.

Tohsato, Y., Ho, K.H.L., Kyoda, K., and Onami, S. SSBD: A database of quantitative data of spatiotemporal dynamics of biological phenomena. *Bioinformatics* 2016. academic.oup.com. https://academic.oup.com/bioinformatics/article-abstract/32/22/3471/2525594.

Williams, E., Moore, J., Li, S.W., Rustici, G., Tarkowska, A., Chessel, A., et al. 2017. The Image Data Resource: A Bioimage Data Integration and Publication Platform. *Nature Methods* 14 (8): 775–81.

13

The Future of CLEM: Summary

Lucy Collinson[1] and Paul Verkade[2]

[1] The Francis Crick Institute, London, United Kingdom
[2] University of Bristol, Bristol, United Kingdom

Over the course of this book, we have heard from scientists who are experts in a variety of imaging fields, covering a broad range of biological (and some physical) applications of correlative imaging. We have heard about the current state of the art in the analysis of the structure–function relationship, and though every experimental workflow varies according to the scientific question, the same patterns and trends arise time and again through these different voices. From this, we can understand the potential of correlative imaging (CI) and can define a wish list of strategic and technical developments focused on revealing a deeper understanding of biological and physical systems.

At the most fundamental level, there is a call to properly define CI. This definition could be "a workflow in which a single sample is imaged sequentially using different imaging modalities to extract different and complementary information," in contrast to comparative imaging or multimodal imaging, where imaging is used as a locator tool, or different samples of the same type are imaged in different modalities and compared. Here, Lippens and Jokitalo and Anderson, Nillson, and Fernandez Rodriguez make an important point, that where comparative or multimodal imaging is sufficient to answer the scientific question, time and resources can be saved by forgoing more complex correlative workflows. For those new to the field, much can be learnt from the examples in this book where correlative imaging was necessary to answer the scientific question – characterization of actin-rich knobs in adhesion of *Plasmodium falciparum*-infected erythrocytes, the study of membrane bending by ESCRTS, differentiation of COP-coated vesicles in subcellular trafficking pathways, and changes in synapses in response to learning or disease.

Where correlative imaging is deemed necessary, it is important to define the core reason. In some cases, correlative imaging is used as a way to track or target a rare event (e.g. a tumor cell migrating through a blood vessel wall), the ultrastructure of which is then analyzed for scientific meaning. Targeted imaging of this type has associated benefits in reduced data size and acquisition time. In other cases, it is necessary to detect functional markers in the context of sample structure to understand their function (usually highlighting the location of macromolecules in cells or tissues in biological applications), which often demands a higher accuracy correlation than tracking experiments. In both cases, correlative imaging involves compromises, between resolution and field of view, between fluorescence preservation and electron contrast, between speed and volume, between sample size and native-state preservation, between time and resolution, etc. For example, Lafont describes specific compromises in correlation of superresolution light microscopy and atomic force microscopy, in that the laser used for cantilever positioning can cause photobleaching of fluorophores, and the antibody labeling to highlight macromolecules can reduce AFM localization accuracy.

Several challenges must be overcome to make correlative imaging more robust and accessible:

- Minimizing artefacts from radiation damage caused by one imaging modality and visible in the next
- Improved chemistry in the form of dual-modality probes and new purpose-designed embedding resins
- Improved workflows for sample preparation and sample transfer
- New ways of creating landmarks in samples that are visible in different imaging modalities without disrupting or masking information and that enable automated image registration
- New file formats for correlative data that store transformation matrices, warping parameters, ROIs, and aligned image layers
- Development of artificial intelligence and deep learning algorithms for image overlay, smart tracking, and image segmentation
- Application of virtual reality to on-the-fly ROI tracking and data visualization
- Development of integrated imaging tools focused on a set of standard correlative workflows
- End-to-end automation to increase speed and throughput (one of the major drawbacks of CI approaches), leading to democritization of correlative microscopy, uptake by non-specialists, increase in sample throughput, and thus statistical significance and reproducibility

Future work in the field to address these challenges must focus on real and pressing scientific questions that can only be answered using CI. This will make the technology specifications more stringent, but will ultimately deliver solutions that work for real-life applications, and avoid shelving of expensive sub-optimal correlative and integrated technologies.

The experts who contributed to this book pose new ideas for technical innovations in correlative imaging that could enable new science, and outline their wish lists for "perfect" correlative experiments, which help to define the gaps where technical developments would impact the field. Koning, Koster, and Hoogenboom seed the idea of EM as ground-truth for molecular localizations in other imaging modalities. Anderson, Nillson, and Fernandez-Rodriguez suggest adaption of FRET probes for CLEM, to enable detection of active versus inactive populations of macromolecules, making the technique itself a functional imaging tool for understanding molecular interactions. Contributors suggest many other information types that could be correlated with imaging data to reveal new scientific knowledge: chemical analysis (via Raman or TOF-SIMS);

spatial elemental analysis (EELS, EDX, nanoSIMS, OrbiSIMS, XRF etc); nucleic acid detection for spatial gene expression studies; electrophysiology for neuronal communication; and scanning probe microscopies for physical measurements including stiffness, dissipation, elasticity, adhesion, ligand/ receptor interaction forces and cell mechanics. Patwardhan and Swedlow similarly suggest linking imaging ontologies to omics and other ontologies for meta-analyses of correlative data.

Two major scientific fields that could benefit from correlative imaging, at different ends of the size and resolution scales, are structural biology and clinical studies. Bharat and Kukulski speculate that targeted imaging of fluorescently labeled interacting macromolecular complexes and cofactors, or time-resolved imaging of fluorescently labeled membrane fusion events, could lead to collection of data reaching statistical significance on observations with large savings in microscope and reconstruction time. Limitations in the resolution of cryo-sample preparation, cryo-fluorescence microscopy, radiation damage, and probes would need to be overcome.

As noted by Guerin, Liv, and Klumperman, the first "correlative" experiments were performed using biopsies. We are now seeing a growing activity in correlation of clinical samples, to understand subcellular changes and signatures in infected or diseased tissues, and in subcellular fate of targeted drug delivery. For correlative imaging to reveal useful information from clinical samples, a continuum of imaging across scales and different imaging media (sound waves, magnetic fields, X-rays, photons, electrons, probe tips, and ions) without barriers must be developed. Micheva and Kolotuev suggest that correlative imaging could be applied at the human scale to relate intracranial electrophysiology during brain surgeries to synapse-level communication.

Ultimately, the fate of the correlative imaging field will depend on biomedical and physical scientists posing complex questions, teamed with a unique subset of imaging scientists who are able to understand, develop and apply a diverse set of advanced technologies to these key scientific challenges. New training initiatives will be needed to produce such imaging scientists. They will have to build and connect teams of multidisciplinary scientists to perform the work, and a new view from funders will be required to support such systems-level collaborations, with commitment from commercial suppliers to useful translation of correlative solutions.

As noted by Anderson, Nillson, and Fernandez Rodriguez, formalization and growth of the CI community will require networking across disciplines (microscopy multiculturalism), mimicking the networking of imaging technology to deliver a correlative experiment. Most contributors to this book, including Peddie and Scheiber, Micheva and Kolotuev, Paul-Gilloteaux and Schorb, and Patwardhan and Swedlow, call for greater efforts to build shared community resources, to include open repositories for samples, fiducials, algorithms, data and protocols, to help reduce costs, enhance reproducibility, and make open collaboration more commonplace and effective. In the spirit of this uniquely collaborative community, and with the aim of expediting results, open source resources should be pursued wherever possible.

It is encouraging to see that the issues raised in this book are also being recognized and addressed at a higher level. The EU-funded COST project titled "COMULIS" (COrrelated MUltimodal imaging in the LIfe Sciences, www.comulis.eu) started at the end of 2018 and is a concerted effort to bring together the CI communities from the more basic life sciences research and those from (pre-)clinical imaging. The project has set out to define the standards for correlative imaging, address nomenclature, and foster crossover between the two fields of expertise. We can learn so much from each other! Although initiated in Europe and still in its infancy, as we write this, the community is growing rapidly, and we encourage you to engage with the initiative. There are already links with other groupings in the United States and beyond (e.g. Microscopy Australia), and that will surely strengthen the field even more.

Index

Correlative Imaging: Focusing on the Future, First Edition. Edited by Paul Verkade and Lucy Collinson.
© 2020 John Wiley & Sons Ltd. Published 2020 by John Wiley & Sons Ltd.